Examining the Impact of Water Quality on the Integrity of Distribution Infrastructure

Awwa
Research
Foundation

About the Awwa Research Foundation

The Awwa Research Foundation (AwwaRF) is a member-supported, international, nonprofit organization that sponsors research to enable water utilities, public health agencies, and other professionals to provide safe and affordable drinking water to consumers.

The Foundation's mission is to advance the science of water to improve the quality of life. To achieve this mission, the Foundation sponsors studies on all aspects of drinking water, including supply and resources, treatment, monitoring and analysis, distribution, management, and health effects. Funding for research is provided primarily by subscription payments from approximately 1,000 utilities, consulting firms, and manufacturers in North America and abroad. Additional funding comes from collaborative partnerships with other national and international organizations, allowing for resources to be leveraged, expertise to be shared, and broad-based knowledge to be developed and disseminated. Government funding serves as a third source of research dollars.

From its headquarters in Denver, Colorado, the Foundation's staff directs and supports the efforts of more than 800 volunteers who serve on the board of trustees and various committees. These volunteers represent many facets of the water industry, and contribute their expertise to select and monitor research studies that benefit the entire drinking water community.

The results of research are disseminated through a number of channels, including reports, the Web site, conferences, and periodicals.

For subscribers, the Foundation serves as a cooperative program in which water suppliers unite to pool their resources. By applying Foundation research findings, these water suppliers can save substantial costs and stay on the leading edge of drinking water science and technology. Since its inception, AwwaRF has supplied the water community with more than $300 million in applied research.

More information about the Foundation and how to become a subscriber is available on the Web at **www.awwarf.org.**

Examining the Impact of Water Quality on the Integrity of Distribution Infrastructure

Prepared by:
Rehan Sadiq, **Syed A. Imran**, and **Yehuda Kleiner**
Institute for Research in Construction
National Research Council of Canada, 1200 Montreal Rd., Ottawa, ON, K1A 0R6, Canada

Jointly sponsored by:
Awwa Research Foundation
6666 West Quincy Avenue, Denver, CO 80235-3098
and
Institute for Research in Construction
National Research Council of Canada
Ottawa, ON, K1A 0R6, Canada

Published by:

Awwa Research Foundation

American Water Works Association

IWA Publishing

DISCLAIMER

This study was jointly funded by the Awwa Research Foundation (AwwaRF) and the National Research Council of Canada (NRC Canada). AwwaRF and NRC Canada assume no responsibility for the content of the research study reported in this publication or for the opinions or statements of fact expressed in the report. The mention of trade names for commercial products does not represent or imply the approval or endorsement of either NRC Canada or AwwaRF. This report is presented solely for information purposes.

CONTENTS

LIST OF TABLES

LIST OF FIGURES

FOREWORD

The Awwa Research Foundation is a nonprofit corporation that is dedicated to the implementation of a research effort to help utilities respond to regulatory requirements and traditional high-priority concerns of the industry. The research agenda is developed through a process of consultation with subscribers and drinking water professionals. Under the umbrella of a Strategic Research Plan, the Research Advisory Council prioritizes the suggested projects based upon current and future needs, applicability, and past work; the recommendations are forwarded to the Board of Trustees for final selection. The foundation also sponsors research projects through the unsolicited proposal process; the Collaborative Research, Research Applications, and Tailored Collaboration programs; and various joint research efforts with organizations such as the U.S. Environmental Protection Agency, the U.S. Bureau of Reclamation, and the Association of California Water Agencies.

This publication is a result of one of these sponsored studies, and it is hoped that its findings will be applied in communities throughout the world. The following report serves not only as a means of communicating the results of the water industry's centralized research program but also as a tool to enlist the further support of the nonmember utilities and individuals.

Projects are managed closely from their inception to the final report by the foundation's staff and large cadre of volunteers who willingly contribute their time and expertise. The foundation serves a planning and management function and awards contracts to other institutions such as water utilities, universities, and engineering firms. The funding for this research effort comes primarily from the Subscription Program, through which water utilities subscribe to the research program and make an annual payment proportionate to the volume of water they deliver and consultants and manufacturers subscribe based on their annual billings. The program offers a cost-effective and fair method for funding research in the public interest.

A broad spectrum of water supply issues is addressed by the foundation's research agenda: resources, treatment and operations, distribution and storage, water quality and analysis, toxicology, economics, and management. The ultimate purpose of the coordinated effort is to assist water suppliers to provide the highest possible quality of water economically and reliably. The true benefits are realized when the results are implemented at the utility level. The foundation's trustees are pleased to offer this publication as a contribution toward that end.

David E. Rager
Chair, Board of Trustees
Awwa Research Foundation

Robert C. Renner, P.E.
Executive Director
Awwa Research Foundation

ACKNOWLEDGMENTS

The authors of this report are indebted to the following organizations and individuals who contributed to this project with their data, experience, insight and good advice:

AwwaRF officials and project managers:

- Jian Zhang (AwwaRF Project Manager)

Project Advisory Committee (PAC):

- Yakir Hasit – CH2MHILL
- Tahir Husain – Memorial University of Newfoundland
- John D. Keith – Lincoln Water Services

Workshop Participants:

- Balvant Rajani – The NRC Institute for Research in Construction Canada
- David Hubble – The NRC Center for Sustainable Infrastructure Research Canada
- David Hughes – American Water Works Company
- Dunling Wang – The NRC Center for Sustainable Infrastructure Research Canada
- Edward McBean – University of Guelph
- France Lemieux – Health Canada
- Graham Gagnon – Dalhousie University
- Haseen Khan – Department of Environment and Conservation, Newfoundland
- Ian Douglas – City of Ottawa
- Jian Zhang – American Water Works Association Research Foundation
- Kenneth Rotert – United States Environmental Protection Agency
- Manuel Rodriguez – University of Laval
- Marie-Claude Bessner – Ecole Politechnic de Montreal
- Md. Maruf Mortula – The NRC Center for Sustainable Infrastructure Research Canada
- Mikhael Tobalt – NRC Infraguide
- Rehan Sadiq – The NRC Institute for Research in Construction Canada
- Syed A. Imran – The NRC Center for Sustainable Infrastructure Research Canada
- Tahir Husain – Memorial University of Newfoundland
- Thomas Rockaway – University of Louisville
- Yakir Hasit – CH2MHILL
- Yehuda Kleiner – The NRC Institute for Research in Construction Canada

Special thanks to the following individuals who contributed data and advice without any formal commitment:

- Amir A. Khan – Department of Environment and Conservation, Newfoundland
- Charlotte D. Smith – Charlotte Smith & Associates, Inc.
- Christine A. Owens – Tampa Bay Water
- Ginasiyo Mutoti – Timmons Group
- Karen Cossit – The NRC Center for Sustainable Infrastructure Research Canada
- Suibing Liu – CDM

EXECUTIVE SUMMARY

Utilities in the United States increasingly report that their treatment system upgrades are driven by the need to comply with new and forthcoming United States Environmental Protection Agency (USEPA) regulations (AWWA 2000). These treatment adjustments lead to corresponding changes in the distributed water quality. The main objective of this research project is to identify research needs for evaluating water quality impacts on distribution infrastructure integrity. This is achieved by identifying missing links in the complex interrelation among water quality regulations, which drive new treatment technologies, which result in primary and secondary impacts on water quality in distribution system, which ultimately affect the distribution infrastructure integrity.

RESEARCH OBJECTIVES

Four major objectives were specified for this research:

- review of current and proposed USEPA drinking water quality regulations,
- review of common treatment technologies (and practices) for compliance and the associated impact on quality of water distributed,
- review of impact of water quality on common distribution materials, and
- identify research needs and knowledge gaps to establish links between changing water quality regulations, and the impact of compliance strategies on the integrity of the distribution system.

BACKGROUND

State of Practice

Utilities today are under pressure to provide safe drinking water to consumers in acceptable quantity and quality. Changes in regulatory restrictions on source-water withdrawals, treatment techniques and maximum contaminant levels (MCLs) require utilities to constantly change, upgrade or altogether replace their historic water sources and treatment practices. Strategies to comply with the state and national regulations on the drinking water quality and quantity ultimately lead to changes in the chemistry (quality) of the water that comes in contact with the distribution networks. Though there is a general awareness that these changes might adversely impact the distribution infrastructure, utilities have historically taken a myopic outlook on protection of distribution infrastructure due to impacts of water quality.

USEPA Drinking Water Regulations

As knowledge and understanding of the impacts of water quality on human health are advancing, USEPA's regulations are becoming increasingly complex and stringent. Promulgations of the Disinfectant/Disinfection Byproduct Rules (D/DBPRs), Surface Water Treatment Rules (SWTRs), Groundwater Rule (GWR), Total Coliform Rule (TCR), and Lead and Copper Rule (LCR) indicate that the USEPA increasingly favors a managed approach to drinking water safety over an MCL at the distribution 'point of entry' approach. However, despite evidence that distribution system deficiencies are among the major causes leading to waterborne outbreaks, a comprehensive regulatory approach to distribution integrity has not been attempted.

Any managed attempts at controlling contaminants (including pathogens) are as good as the state of the distribution system. Even complete compliance with all the regulations does not guarantee safe drinking water unless the distribution infrastructure is in acceptable condition. It is expected that the continuing revisions to the TCR and the proposed Distribution System Rule will address some of the issues arising due to deterioration of distribution systems.

Water Treatment Technologies

The primary objective of selecting water treatment technologies is to produce safe drinking water. However, utilities must also consider needs of present and future expansion, local and national environmental restrictions on siting, source-water usage and disposal of process residuals. Utilities need to pay particular attention to the placement or sequence of the additional treatment units within the existing layout. Secondary water quality changes should also be evaluated to determine potential problems related to adverse impacts on distribution infrastructure.

Distribution Infrastructure Impacts

'Integrity' of distribution infrastructure can be defined as its ability to transport water in acceptable quantity and quality and with minimal interruption.

Distribution infrastructure is a complex network of pipes and appurtenances constructed of different materials, of different age and different manufacturing processes. Changes in water quality can potentially impact the distribution infrastructure by causing deterioration of the internal surfaces.

Just as water chemistry can affect the inner surfaces of the distribution system, the chemical properties of these inner surfaces can affect water chemistry. This circular affects are often not completely understood and are difficult to de-couple. While a plethora of literature exists on the water quality deterioration due to the interaction with internal pipe surfaces, the coupled phenomenon of deterioration of the internal pipe surfaces due to the water quality has only received cursory attention.

Research Priorities for Different Distribution System Materials

A workshop was conducted on 30th March 2006 at the National Research Council Canada – Institute for Research in Construction (NRCC-IRC) campus in Ottawa, Canada. Experts in the field were invited to identify research needs based on a draft synthesis document that was distributed ahead of time. Input from experts was also solicited in the form of a questionnaire survey. Research issues, needs and gaps were categorized as *high*, *medium* and *low* priority, based on the discussion and consensus achieved in the Workshop. Based on the suggestion of the participants, another category relating to the prioritization of the impact of distribution infrastructure on water quality was also included. The insights gained from this project would help in defining future research projects to address identified gaps in knowledge and understanding, while considering water quality as well as impact on the deterioration of distribution infrastructure.

Table E-1
Research priorities identified for unlined iron in the distribution infrastructure

Priority	Impacts on Distribution Infrastructure	Impacts on Water Quality
HIGH	*† Develop and test methods of evaluating the impacts of different disinfectants on deterioration of old pipes. § Control of corrosive action of common disinfectants. *† Identify interactions between corrosion rate, microbiological activity and disinfectant type and residual.	‡ Develop and test methods for comparison of disinfection efficacy for different pathogen groups using different disinfectants. *† Develop models for residual disinfectant depletion due to interaction with corrosion byproducts and pipe walls. ‡ Model and evaluate accumulation, growth and release of physical, chemical and microbiological contaminants.
MEDIUM	‡ Document and model the genesis and evolution of corrosion scale and release controlling solid phases under different water qualities. * Develop guidance for mitigation of internal deterioration in old pipes. § Document and compare performance of newly lined iron pipes on remaining life of pipe.	*† Develop practical solutions for water quality deterioration.
LOW	‡ Develop a forensic model to identify internal deterioration problems based on the type and concentration of the particulate released. ‡ Evaluate accumulation and release of heavy metals from corrosion scales.	§ Evaluate control measures to increase the formation of passivating solid phases.

* Some research needed; ‡ Considerable research done; § No research done yet; † continuing research in active AwwaRF projects

Table E-1 gives the research needs identified for unlined iron pipes. This category includes old galvanized pipes and iron pipes with insufficient lining. Areas of current concern in unlined iron pipes include the accumulation and release of heavy metals like arsenic, lead, cadmium etc., from the internal surfaces. There is also a need for an assessment tool for measuring the internal condition of old pipes. This would help in planning rehabilitation activities.

Table E-2 gives the research needs identified for the lead and copper pipes in the distribution system and household plumbings. For lead and copper pipes a major area of concern is the galvanic corrosion of these materials at the interface where they are in contact with a newer pipe material. For copper pipes, there is an additional need to evaluate whether passivated old copper pipes and newly installed copper pipes exhibit any significant galvanic corrosion.

Table E-2
Research priorities for Lead and Copper in the distribution infrastructure

Priority	Impacts on Distribution Infrastructure	Impacts on Water Quality
HIGH	*† Develop and test methods of evaluating the impacts of different disinfectants on deterioration of copper pipes. § Control of corrosive action of common disinfectants. ‡ Identify interactions between corrosion rate, microbiological activity and disinfectant type and residual.	‡ Develop models for residual disinfectant depletion due to interaction with pipe walls. ‡ Evaluate impacts of chloramines on lead and copper release and LCR compliance. ‡ Evaluate impact of new copper plumbing installations on LCR compliance. ‡ Evaluate impact of partial replacement of lead pipes on LCR compliance.
MEDIUM	‡ Document and model the genesis and evolution of the corrosion scales and release controlling solid phases for lead and copper under different water qualities. ‡ Document and compare corrosion of newly installed and old copper pipes.	* Develop practical solutions for water quality deterioration in both lead and copper pipes.
LOW	§ Develop a forensic model to identify internal deterioration problems based on the type and concentration of the particulate released from both lead and copper pipes.	§ Evaluate control measures to increase the formation of passivating solid phases in both lead and copper pipes.

* Some research needed; ‡ Considerable research done; § No research done yet; † continuing research in active AwwaRF projects

PVC, PE and other plastic components represent a newer and continuously evolving material used in the distribution systems. These materials have a limited history and therefore need further research. Specifically, as shown in Table E-3, the areas of concern are the leaching of different organic components into the drinking water and the degradation of these materials through oxidants (mainly disinfectants) in the drinking water.

Table E-4 gives the research needs for cement pipes. There is significant literature on the impact of ionic components on the corrosion and degradation of concrete and cement structures. However, the degradation of these structures at the low ionic contents typical of drinking water need to be further studied.

Table E-3
Research priorities for PVC, PE and other plastics in the distribution infrastructure

Priority	Impacts on Distribution Infrastructure	Impacts on Water Quality
HIGH	‡ Identify and control permeation of organic contaminants through pipe walls of plastic pipes. †‡ Impact of changing disinfectants on elastomers and other plastic components of the distribution infrastructure.	§ Evaluate impacts of chloramines on leaching from pipe matrix. ‡ Characterization of contaminants leaching from plastic pipes and their impact on health and aesthetic water quality.
MEDIUM	‡ Examine and quantify the leaching of monomers and other organic components from the pipe matrix.	‡ Develop practical solutions for water quality deterioration in plastic pipes.
LOW		‡ Evaluate impacts of new plastic installations on health and aesthetic water quality deterioration.

* Some research needed; ‡ Considerable research done; § No research done yet; † continuing research in active AwwaRF projects

Table E-4
Research priorities for AC, Concrete and other cementatious materials in the distribution infrastructure

Priority	Impacts on Distribution Infrastructure	Impacts on Water Quality
HIGH		‡ Model and evaluate accumulation and release of heavy metals from pipe surfaces.
MEDIUM	‡ Evaluate softening of pipe materials under different water quality scenarios. § Evaluate impact of disinfectants on integrity of pipe.	‡ Evaluate impact of new installations on health and aesthetic water quality deterioration.
LOW	*† Corrosion of reinforcing steel or prestressing wires and impact on integrity of concrete structures and pipes.	

* Some research needed; ‡ Considerable research done; § No research done yet; † continuing research in active AwwaRF projects

Table E-5
Research priorities for liners (polyaromatic hydrocarbons, epoxy, etc.) in the distribution infrastructure

Priority	Impacts on Distribution Infrastructure	Impacts on Water Quality
HIGH	‡ Evaluate the long-term performance of lined pipes and liners.	
MEDIUM	‡ Leaching of PAHs, monomers and other matrix components from liners and their impact on the durability and effectiveness of linings.	‡ Evaluate impact of new installations on health and aesthetic water quality deterioration. ‡ Evaluate the long-term leaching of contaminants from liners.
LOW	‡ Impact of disinfectants on performance of liners.	

* Some research needed; ‡ Considerable research done; § No research done yet; † continuing research in active AwwaRF projects

Table E-5 gives some of the research priorities identified for lining material in the distribution systems. These liners are typically used either as a component of new pipes or for rehabilitation of old pipes. In both cases, the long-term performance of the liners needs to be evaluated.

Other research priorities identified in Table E-6 address water sources, water treatment and operation, maintenance and renewal activities, and are intended to minimize disruption to the existing distribution infrastructure. There is also an urgent need for addressing scale-up issues in laboratory and bench-scale experiments as they apply to the distribution system.

Different utilities have different research needs based on the applicability of specific regulations, and use of different treatment processes and materials of constructions. The authors have proposed an innovative but simple framework (Hierarchical Relational Model) by which utilities can evaluate the impact of changes in treatment processes to the existing infrastructure. Different materials have different, often opposing water quality requirements to mitigate deterioration. Conflicting water quality impacts on different distribution materials can also be identified using this model. The model proposed in this project is intended as a proof of concept and can be further refined to incorporate more complex real systems. Further development of similar assessment models to aid decision-making in evaluating treatment changes as well as evaluating material for the replacement of existing infrastructure is needed.

Table E-6
Other research priorities for distribution infrastructures

Priority	Impacts on Distribution Infrastructure	Impacts on Water Quality
HIGH	‡ Develop and evaluate methods to identify water quality impacts on different materials simultaneously. ‡ Develop a standard assessment procedure to identify the extent of internal deterioration in pipes and other inaccessible infrastructure. ‡ Develop methods to evaluate scale-up issues arising from bench and laboratory studies.	‡ Develop methods to evaluate and optimize water sources and treatment processes to protect the existing distribution infrastructure.
MEDIUM	‡ Determining optimum monitoring strategies to evaluate internal deterioration in distribution systems. ‡ Mechanisms of inhibition by corrosion inhibitors.	‡ Develop methods to evaluate scale-up issues arising from bench and laboratory studies.
LOW	‡ Impact of disinfectants on performance of liners.	

* Some research done – more needed; † continuing research in active AwwaRF projects; ‡ Limited research done; § No research done yet

RECOMMENDATIONS FOR FUTURE RESEARCH

General recommendations for future research can be summarized as following:

1. The evaluation of how changes in water sources, involving changes in water chemistry, can affect different components of the distribution network. For instance, some surface waters can corrode unlined iron pipes faster than groundwater.
2. The reconciliation of conflicting impacts of chemical properties of water on different pipe materials. For instance, water composition compatible with iron pipes may not be compatible with copper pipes. Therefore, the addition of chemicals to protect iron pipes may cause failure of copper pipes.
3. The evaluation of unintended effects of change in water treatment processes. For instance, increasing the amount of chlorine to kill pathogenic microorganisms can increase deterioration rate in some pipe materials.
4. The evaluation of the effects of frequent changes in water quality on the distribution pipes. For example, evaluating the impacts of the frequency and duration of alternating conditions on distribution pipes when a utility regularly switches between surface water and groundwater sources to meet seasonal demand.
5. The development of a decision-support model (like HRM) to help water utilities evaluate the effect of changing water treatment technologies (practices) on the deterioration of different pipe materials.

CHAPTER 1
INTRODUCTION

Water traveling through 'aging' distribution infrastructure decreases in quality due to the uptake of contaminants through intrusion from compromised components, permeation, leaching and internal corrosion. Though considerable research exists that focuses on the deterioration of water quality through distribution infrastructure, but the converse phenomenon of deterioration of distribution infrastructure has not been studied extensively. This project provides a synthesis document outlining the impacts of changing water quality, due to regulatory drivers, on distribution infrastructure integrity. This task is accomplished in the following four steps:

- review of current and upcoming drinking water quality regulations as drivers of change in the quality of the distributed water,
- review of treatment technologies for compliance and the associated secondary impacts on water chemistry (quality),
- review of the impacts of water chemistry on common distribution infrastructure materials, and
- identify research needs, through an expert workshop, to establish links between the changing water quality regulations and compliance strategies on distribution integrity.

The main focus of this study is to identify how the water distribution network might be impacted by water quality changes that occur when water utilities attempt to comply with new regulations. The research plan involves the development of a synthesis document outlining major changes in water quality (as a result of changes in treatment techniques for regulated contaminant control) and its impact on major components of the distribution system.

In this report, the terms 'distribution infrastructure', 'water distribution network', 'water distribution system', 'distribution network' and 'distribution system' are used interchangeably to mean 'water distribution infrastructure', including pipes and appurtenances (valves, fittings).

THE INTEGRITY OF WATER DISTRIBUTION INFRASTRUCTURE

In this report, 'integrity' of distribution infrastructure is defined as its ability to transport water in acceptable quantity and quality without causing any 'structural' or 'functional' failure of its components.

Water quality failures observed in distribution system are often an indicator of the long-term deterioration of the distribution infrastructure. For instance, finished water introduced into the distribution system might cause metallic corrosion and lead to red water problems at the consumers tap. From a public water system (PWS) point of view, a water quality change (increase in color) is often the first indication that the integrity of the distribution system is being compromised in some way. A PWS may conduct further investigations to ascertain the cause of water quality deterioration.

Though the 'structural' integrity of the system may be intact, the 'functional' integrity could be compromised because of the failure of the distribution system to avoid deterioration of the transmitted water. In actual PWS experience, this functional failure may persist for years before leading to actual structural failure. For instance, water quality problems related to distribution system such as red water, loss of residual and subsequent microbiological proliferation may persist for years before actual failure of the pipe. Thus the water quality at the consumers tap can be used as an indicator of the "health" of the distribution infrastructure. Since distribution infrastructure is essentially a buried structure, these indirect evidences are often used to measure the impact of finished water quality on its integrity.

PWSs have traditionally taken a long-term approach to maintaining and rehabilitating the distribution infrastructure when compared with the short-term urgency of water quality deterioration. Measures such as distribution system flushing and shock (booster) chlorination may provide short-term relief without solving the underlying long-term problems related to deterioration of distribution components.

Compromise on the integrity of water distribution infrastructure can be caused by a number of water quality deterioration mechanisms including internal corrosion, leaching, contaminant intrusion, microbial regrowth and permeation of organic compounds through system components (Sadiq, Kleiner, and Rajani 2004). Failure at the distribution level can be critical because it is closest to the point of use and there are virtually no safety barriers before consumption. PWSs must manage the source-waters, treatment processes and distribution infrastructure holistically in order to maintain the quality at the consumers tap.

DRINKING WATER REGULATIONS AND WATER QUALITY

PWSs are under tremendous pressure to meet national and state regulations/ guidelines for maintaining maximum contaminant levels (MCLs) in the distribution system, in order to provide consumers with safe drinking water. The United States Environmental Protection Agency (USEPA) established national drinking water regulations in the US, under the Safe Drinking Water Act (SDWA) of 1974. Currently 91 contaminants are regulated by establishing MCLs in the treated water before introduction into the distribution system (Pontius 2004).

Traditionally, utilities have managed the mandated compliance levels by upgrading and optimizing their treatment processes or by changing source-waters (Daniel 1998, Taylor et al. 2005). Recent and proposed future regulations are more complex and favor a managed, multiple strategy approach to maintaining safe drinking water quality. For instance, a surface water system is expected to reduce overall risk by balancing the short-term (acute) and certain microbiological risk with the potential long-term (chronic) and uncertain risk from disinfection byproducts (DBPs). The USEPA addresses this *risk-risk* trade-off by promulgating paired regulations, Stage 1 Disinfectant/Disinfection Byproduct Rule and Long Term 1 Enhanced Surface Water Treatment Rule (Stage 1 D/DBPR and LT 1 ESWTR), simultaneously. Specific provisions in these paired regulations require utilities to comply with both the regulations. This encourages utilities to carefully evaluate treatment alternatives to reduce DBPs while maintaining microbial protection (Roberson 2003). The Groundwater Rule (GWR) and Stage 2 D/DBPR are expected to provide similar provisions for groundwater systems.

Changes in the treatment techniques to reduce or eliminate any contaminant of concern invariably lead to changes in the final water quality. When this changed water quality is introduced into the distribution system it may trigger a cycle of changes, in which the distribution system as well as the final water quality are likely to be impacted (Imran et al. 2005b).

For instance, one way of maintaining simultaneous compliance with the paired regulation is to change the secondary disinfectant from chlorine to chloramine. Chloramine is more stable in the distribution system and can therefore maintain a residual longer than chlorine. An added benefit to utilities is that DBPs formation by chloramine is significantly lower than that for chlorine. However, there are some specific problems associated with chloramine like lower potency for microbial inactivation compared to chlorine, nitrification (Liu et al. 2005) and increased potential for lead release (Vasquez et al. 2006).

Therefore, a change in disinfectant to comply with the Stage 1 D/DBPR and LT1ESWTR could lead to a potential violation of the Lead and Copper Rule (LCR), the Total Coliform Rule (TCR) or the MCL for nitrates and nitrites (USEPA 1999b). The corresponding distribution impact would be increased pipe breakages due to corrosion of lead service lines or dissolution of lead from solders and alloys. Recent research also points to the deterioration of elastomers (gaskets) in distribution fittings (Bonds 2004). Therefore, it is necessary that while evaluating different compliance strategies, the impacts on the existing distribution infrastructure are given due consideration. However, the historical focus of compliance has been on the water quality at the point of entry to the distribution system and the impact of water quality on the distribution system has largely been ignored.

DETERIORATION OF THE DISTRIBUTION INFRASTRUCTURE

It is estimated that there are 880,000 miles of drinking water distribution systems in the US, of which 4,400 miles are replaced annually (Kirmeyer, Richards, and Smith 1994). Based on the estimated replacement rate of about 4,400 miles per year, the average age of the distribution system before it is replaced would be at least 200 years. Grigg (2005) reports that though pipe inventory has gone up by 11% in the period 1992 - 2002, replacement of old pipes seemed to have slowed down, mainly due to increased costs and absence of direct regulatory controls on pipe integrity.

A study by AWWA (2001) estimated that $250 billion would be required over the subsequent 30 years to replace worn-out drinking water infrastructure. A more conservative estimate by the USEPA (2001b) placed the total drinking water infrastructure need over the next 20 years at $151 billion. Transmission and distribution projects were identified as the largest category of need with $83 billion required over the next 20 years. The cost of new treatment processes and upgrades to current treatment processes, required to meet current and near-future regulations, was estimated at $31 billion.

A typical water distribution network is a complex system. Pipes could be of different materials including cast iron, ductile iron, steel, copper, lead, galvanized steel, reinforced concrete, asbestos cement, thermoplastics including polyvinyl chloride (PVC), chlorinated polyvinyl chloride (CPVC), high density polyethylene (HDPE), polybutadiene and composites such as glass fiber-reinforced plastic (GFRP). In addition to the pipes and water mains, there are a number of ancillary components like coatings, gaskets, o-rings, fittings, valves and solders. Additional variability is introduced by the presence of man-made (cement, epoxy, polymeric or calcite lining) and/or naturally occurring films (corrosion byproducts and biofilms) on the interior of the pipe (Broo, Berghult, and Hedberg 2001b; Tomboulian et al. 2004).

Different components of a distribution system have varying responses to the changes in water quality. Often a change in practice may trigger conflicting results (a positive influence on one component and a negative influence on another). For instance, increasing alkalinity to reduce iron release in distribution systems can cause increased copper release in household plumbing (Imran et al. 2006). Therefore, managing water quality to control the deterioration of all the components may not be practicable.

In this report, different distribution components are identified based on their material composition. This will help avoid a distinction between different components based on the same material. For instance, pipes and tanks might be constructed with cement concrete. Though the overall deterioration of both the structures may be different due to other factors (including hydraulics, detention time, erosion etc), the deterioration due to water quality is expected to be similar due to the similarity of the water-surface interactions. Therefore, water quality impacts on distribution systems are best studied for material of a component in contact with water rather than the geometry or functionality of the distribution system component. A flowchart outlining the steps involved in the development of the synthesis document is given in Figure 1.1.

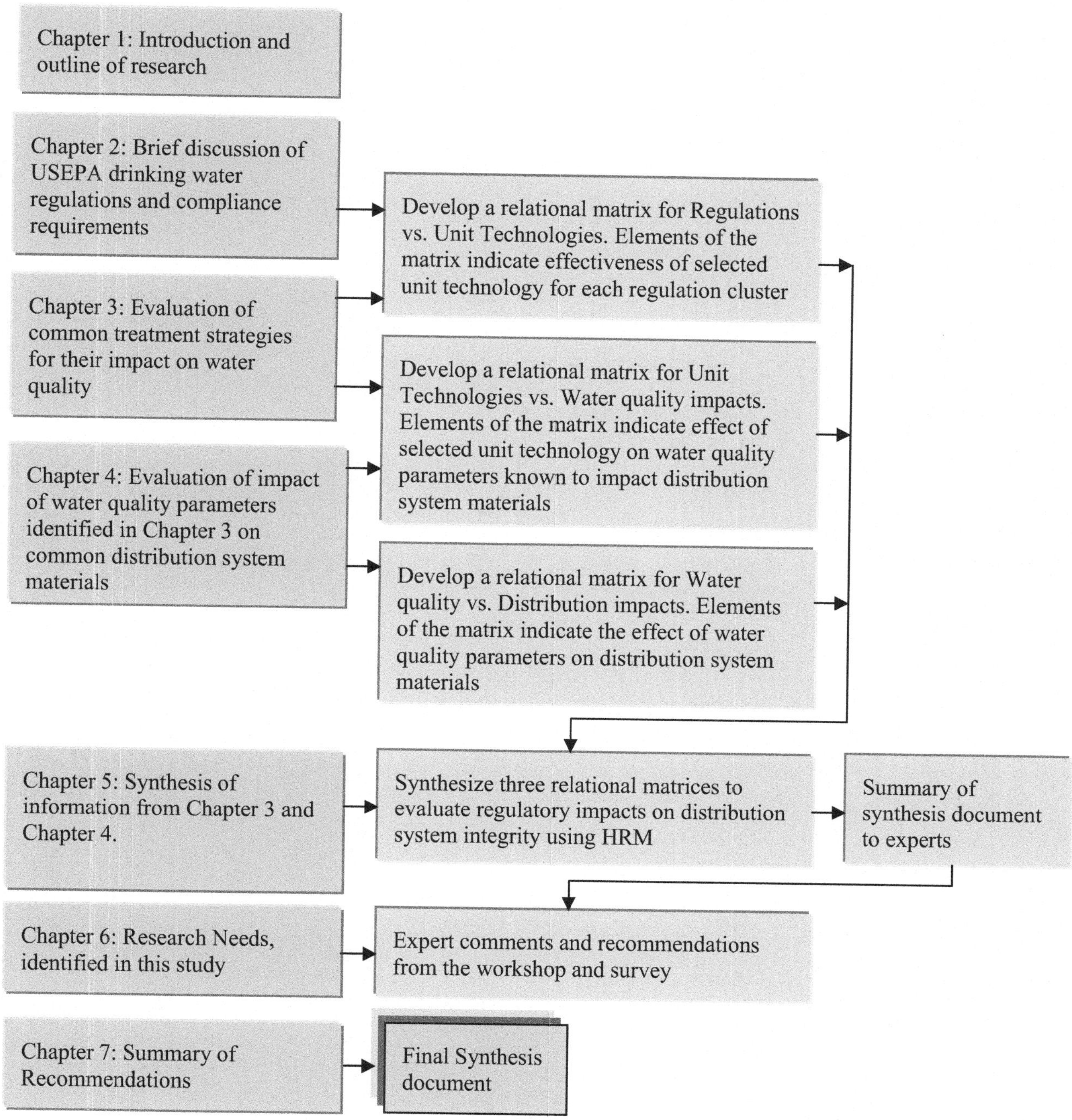

Figure 1.1 Flowchart for development of final synthesis document

CHAPTER 2
REGULATORY DRIVERS FOR CHANGING WATER QUALITY

Drinking water quality standards and regulations define in quantitative terms water that is 'safe' for human consumption by imposing acceptable levels on contaminants in drinking water. This chapter presents a brief overview of the US drinking water regulations and standards to maintain the safety of drinking water.

USEPA DRINKING WATER STANDARDS

The first regulation related to safe drinking water in the US was the abolition of *common cups* on carriers in interstate commerce, promulgated under the 1893 Interstate Quarantine Act in the year 1912. The 1914 Treasury Standards were established, to regulate drinking water quality, by the United States Public Health Service (USPHS), and later revised in 1925, 1942, 1946 and 1962.

Through the 1974 SDWA and its subsequent amendments in 1986 and 1996, the United States Congress authorized USEPA to regulate contaminants that present health risks and which are known to occur in some public drinking water supplies. The USEPA accomplished this task in two steps. The first step involved the creation of the National Interim Primary Drinking Water Regulations (NIPDWRs) in 1975, based largely on the 1962 USPHS standards. The second step involved the revision of these standards following a comprehensive review of the risks posed to water consumers from the regulated contaminants.

The USEPA sets legal limits for the regulated contaminants in drinking water through promulgating MCLs and by suggesting treatment technologies (techniques and practices). The MCLs are intended to be as protective of human health as is "feasible". Feasibility is dependant on the contaminant concentrations, removal efficiencies using best available treatment (BAT) technologies (economically achievable) and the accuracy of analytical measurements. In cases where available analytical methods or normal monitoring schemes are inadequate, the USEPA may impose treatment techniques to be followed under specified conditions. The SDWA gives individual states the authority to set and enforce their own drinking water standards as long as these are at least as stringent as the USEPA's national standards.

The 1986 Amendments to the SDWA required the USEPA to set MCL Goals (MCLGs) and National Primary Drinking Water Regulations (NPDWRs) for 83 specific contaminants. The Amendments included 15 provisions for periodic review of the data and studies affecting the basis of the MCLGs and MCLs. The 1996 Amendments of the SDWA resulted in several provisions related to the Small Systems Compliance Technology Lists. Table 2.1 provides a chronology of the USEPA regulations.

The MCLs are not the only regulatory constraints on a public water service. A number of other checks and balances exist in the form of source-water consumptive use permits, protection of wetlands and drainage areas, process waste discharge permits (Narasimhan 2005), use (restriction of use) of construction material and treatment chemicals etc. For this project, only the regulatory drivers for maintaining safe drinking water are considered.

REGULATED CONTAMINANTS SPECIFIC TO WATER SOURCES

Though all the USEPA drinking water regulations, with MCLs, attempt to mitigate the health risk at the 'point of use', from a utilities perspective some contaminants are comparatively easier to control. These contaminants are removed by conventional technology in place or by slight modification of existing processes (conventional coagulation vs. enhanced coagulation) and do not re-occur after removal. These contaminants are best monitored at the 'point of entry' into the distribution systems. Most organic and inorganic contaminants can be classified under this category. Included in this category are contaminants listed in the Phase I, II and V Rule, the Arsenic Rule and the Revised Radionuclide Rule (Table 2.1). A significant change made by the USEPA is that some contaminants (Radionuclide Rule contaminants) need to be in compliance at all points of entry into the distribution system, where previously only a representative point in the distribution system was monitored for compliance.

Some contaminants can also be classified based on their ease of removal by certain treatment processes. For instance, granular activated carbon (GAC) is effective in removing more than 80% of all regulated organic contaminants from drinking water sources and can be termed as an organic contaminant-specific technology. Similarly, coagulation and filtration can be termed as effective treatment technologies for most heavy metals.

Microorganisms in source waters can present another challenge for compliance due to uncertainty of occurrence and complexity of sampling and quantification. The USEPA attempts to minimize this risk by providing minimum level of treatment for certain classes of microorganisms categorized with respect to its prevalence in certain source waters (groundwater, surface water or groundwater under the direct influence of surface water (GWUDI)). Treatment techniques, complemented by rigorous monitoring, can reduce the endemic and acute health risks associated with pathogenic microorganisms.

Various SWTRs, the TCR and the proposed GWR attempt at mitigating the most endemic and difficult-to-treat pathogenic microorganisms present in many source waters by establishing minimum treatment criteria instead of having treatment goals. This is due to the large variability in the level of microorganisms in the source water due to seasonal variations or during epidemics (Davies et al. 2005). The recent LT 2 ESWTR is the most comprehensive microbiological control regulation for surface water and GWUDI sources where the required level of treatment is linked to the prevalence of pathogenic microorganisms in the source-waters. Reliance on a single strategy of treatment (either filtration or disinfection) is discouraged to provide a more comprehensive safety barrier in case of occasional treatment deficiencies.

When criteria of compliance for different contaminants are in conflict, the USEPA promulgates the conflicting regulations in pairs. For instance, various SWTRs require different levels of treatment for different pathogens and one way of complying would be to disinfect with high disinfectant concentration or provide long contact times. However, this may lead to the problems of either higher exposure levels to disinfectants or to higher DBPs due to long contact times. This conflicting objective is balanced by providing the counter-regulation to the SWTRs in the form of the D/DBPRs where the disinfectant and disinfectant byproducts are regulated.

Table 2.1
Summary of USEPA rulemaking for drinking water contaminants under SDWA

Year	Final Regulations	Federal Register Notices
1975-76	NIPDWRs	40(248):59566-59588 41(133):28402-28409
1979	TTHMs NSDWR	44(231);68624-68707 44(140):42195-42202
1986	Fluoride rule	51(63):11396-11412
1987	Phase I	52(74):12876-12883
1989	TCR SWTR	54(124):27544-27568 54(124):27486-27541
1991	Phase II (a,b) LCR	56(20):3526-3597 56(126):30266-30281 56(110):26460-26564
1992	Phase V	57(138):31776-31849
1995		
1998	Stage 1 D/DBPR Interim ESWTR	63(241):69390-69476 63(241):69478-69521
2000	Radionuclide	65(236):76708-76753
2001	Arsenic	66(14):6976-7066
2002	LT 1 ESWTR	67(9):1812 – 1844
Proposed	GWR	65(91):30193-30274
Proposed	Radon Rule	64(211):59245-59254
2006	LT 2 ESWTR	68(154):47639-47688
2006	Stage 2 D/DBPR	68(159):49547-49596

D/DBPR – Disinfectant/Disinfection Byproduct Rule, ESWTR - Enhanced Surface Water Treatment Rule, GWR – Groundwater Rule, LCR – Lead and Copper Rule, LT 1 ESWTR – Long Term 1 ESWTR, LT 2 ESWTR – Long Term 2 ESWTR, NIPDWR – National Interim Primary Drinking Water Regulation, SWTR – Surface Water Treatment Rule, TCR – Total Coliform Rule, TTHM – Total trihalomethanes

Table 2.2 gives some of the regulations that might require a change in treatment practice or source waters for a non-compliant water system. It is noted that depending on their size and water source, PWS may not have to comply with all the drinking water regulations. Most regulations are limited to PWSs of particular size and source. Therefore, where an effort has been made to include all the relevant drinking water regulations, water purveyors should determine the regulations applicable to them based on their size and water source.

CONTAMINANTS GENERATED WITHIN THE WATER DISTRIBUTION INFRASTRUCTURE

Certain contaminants are generated within the distribution system due to the interaction of bulk water constituents or the interaction of bulk water with pipe surfaces. Trihalomethanes (most common DBPs) formation is an instance where bulk water constituents (organic matter and disinfectant) interact to form undesirable byproducts. The DBPs of chlorine were regulated under the Total Trihalomethanes NIPDWRs. As utilities began using alternate disinfectants; other potentially harmful DBPs associated with ozone and chlorine dioxide came to be recognized and regulated under the Stage1 D/DBPR and the recent Stage 2 D/DBPR.

Residual disinfectant in the distribution system presents a unique challenge for both regulatory agencies as well as the PWSs. Though residual disinfectant may be required for control of microorganisms in the distribution system, a high concentration may render the water unacceptable for human consumption. The health problems associated with high residual disinfectant are mitigated under the D/DBPRs by regulating the maximum concentration of chlorine, chloramines, ozone and chlorine dioxide in the distributed water. In contrast, the regulation (TCR) dealing with microbiological control in the distribution system requires a minimum (detectable) residual disinfectant concentration at all points within the distribution systems.

The interaction of the bulk water with the pipe surfaces may lead to corrosion and leaching of pipe material into the transported water. Lead and copper corrosion, biofilm sloughing and red water problems are instances where the bulk water interacts with the pipe surfaces. The LCR is unique among the USEPA rules in that it regulates the contaminants that are derived based on the corrosion of the distribution system. Therefore, monitoring is done at the consumers' tap and not at the treatment plant. If 10% of the samples exceed the action levels for lead (15 μg/L) and copper (1.3 mg/L), the utility has to take documented action to reduce the lead and copper levels. The LCR monitoring requirements presents a challenge to utilities in determining the location and the frequency of sampling as well as collaborating with the consumers to get the samples at their taps. However, the LCR is very difficult to enforce due to diverse age, lengths, residence times, locations, state of corrosion and materials of household plumbing. Some of these variations are mitigated by following a standard sampling protocol, while others are inevitable due to the complex nature of the distribution network.

REGULATORY SHORTFALLS

Out of more than 14 thousand reported violations for health-based standards in 2002, the TCR violations (69%) far exceeded the chemical contaminants based violations (10%), the LCR violations (5%) or the SWTR violations (15%) (USEPA 2004). The violations must also be viewed in terms of applicability of regulations. The TCR applies to all water systems, while the

LCR and SWTR are limited by the size of the system classification or type of source water. The higher incidence of TCR violations underlines the importance of maintaining distribution infrastructure in a good condition (Roberson 2003).

Distribution system deficiencies in community water systems caused 31% of the reported waterborne outbreaks during 1971-2000. During the same period, distribution system contamination was responsible for 8% of the outbreaks in non-community water systems. The remaining outbreaks were attributed to treatment breakthrough and disinfection deficiencies (Craun, Calderon, and Craun 2003). Contaminants entered in the distribution system through cross-connections, back-siphonage, corrosion and leaching of metals, broken or leaking mains, storage deficiencies and construction and repair of mains. Craun, Calderon, and Craun (2003) reported that the most important chemical contaminant causing distribution system outbreaks was copper. Acute outbreaks associated with copper plumbing were primarily due to corrosion and plumbing deficiencies.

Iron in drinking water networks arises from two sources; breakthrough from the treatment plants and corrosion and subsequent release of corrosion byproducts from iron pipes. The aesthetics of the water deteriorate at a much lower iron level than its health related maximum level. Therefore, iron levels are managed under the National Secondary Drinking Water Regulations (NSDWRs). Some states and utilities have adopted the NSDWRs as mandatory water quality goals. However, the usual point of monitoring for iron is at the point of entry to the distribution system and does not consider the iron contributed by the distribution system itself. Iron contributed from iron pipes is a major source of turbidity and colored water within the distribution system and has been linked to increased disinfectant demand, microbiological-shielding from disinfectant action and in microbiological proliferation.

With one notable exception of the LCR, the USEPA does not require the water quality to be protective of the distribution system. Other regulations like the TCR might require integrity of distribution systems for compliance (intrusion through compromised distribution infrastructure may cause microbiological failures). This compliance is ultimately achieved through water quality criterion (residual chlorine) and by infrastructure rehabilitation.

Utilities traditionally have relied on single source-waters and therefore had to deal with minimum variations in water quality and in the type or concentration of contaminants. This practice afforded the utilities to establish a single compliance strategy for the specific regulated contaminants in their source water. However, over-reliance on single sources is being abandoned due to adverse environmental impacts associated with unsustainable withdrawals. Today, utilities (especially in communities with high growth) may have multiple source-waters and hence complex compliance strategies. For instance, a utility having separate treatment of groundwater and surface water (and complies with regulations individually), does not have a guidance on how mixing of the waters within the distribution system and its related problems will affect its compliance with regulations. Such mixing may have a complex impact on the pH, DBP precursors and formation potentials, and corrosion related problems in distribution systems.

Since the primary objective of the drinking water regulations is the protection of human health, distribution system deficiencies should be taken seriously or even regulated. As stated earlier, the first drinking water regulation in the US was the abolition of *common cups* on interstate carriers. The spirit of this regulation that "*the quality of drinking water is as good as the container in which it is served*" should be followed in the case of distribution systems as well.

Table 2.2
Drinking water regulatory requirements that might require change in water source or treatment technique for compliance

	Surface Water Systems and GWUDI				Groundwater Systems
Population Served	25 – 500	501 – 3,300	3,301 – 10,000	> 10,000	All
Inorganic contaminants (includes arsenic)					
• Compliance with MCLs	•	•	•	•	•
Organic contaminants					
• Compliance with MCLs	•	•	•	•	•
Total Coliform Rule					
• Compliance with MCLs	•	•	•	•	•
SWTR					
• Turbidity performance criteria	•	•	•	•	
• Giardia and virus log removals	•	•	•	•	
• Disinfection level	•	•	•	•	
• Disinfectant residual at distribution system entry	•	•	•	•	
• Contact time (unfiltered systems)	•	•	•	•	
Interim ESWTR					
• Turbidity performance criteria				•	
• Cryptosporidium log removals				•	
Long term 1 ESWTR					
• Turbidity performance criteria	•	•	•		
• Cryptosporidium log removals	•	•	•		
• Disinfection benchmarking	•	•	•		
Filter Backwash Recycle Rule					
• Recycle requirements	•	•	•	•	
Stage 1 D/DBPR					
• DBP MCLs	•	•	•	•	•
• MRDLs	•	•	•	•	•
• Enhanced coagulation	•	•	•	•	
Lead and Copper Rule					
• Optimal corrosion control	•	•	•	•	•
Radionuclides					
• Compliance with MCLs	•	•	•	•	•
Unregulated Contaminants Monitoring Rule					
CCR Rule					
Security vulnerability assessments					
Stage 2 DBPR					
• Revised DBP MCLs	•	•	•	•	•
Long term 2 ESWTR					
• Monitoring for bin classification	•	•	•	•	
• Application of microbial toolbox measures	•	•	•	•	
• Giardia and virus benchmarking	•	•	•	•	
• Monitoring for rebinning	•	•	•	•	
Radon Rule (proposed)					
• MCL compliance	•	•	•	•	•
Ground Water Rule (proposed)					
• Source monitoring, correction of deficiencies, sanitary survey					•

MCLs – maximum contaminant levels, SWTR – Surface Water Treatment Rule, ESWTR – Enhanced Surface Water Treatment Rules, D/DBPR – Disinfectants/Disinfection By-products Rule, DBPs – disinfection byproducts, GWUDI – Groundwater under the direct influence of Surface waters, MRDLs – maximum residual disinfectant levels, CCR – consumer confidence report

CHAPTER 3
DRINKING WATER TREATMENT

Regulatory compliance with drinking water standards can be achieved by using alternate water sources, treatment chemicals and/or treatment technologies. A survey of 200 PWSs in the US indicated that 50% of the systems anticipated changes in their treatment practices to meet future regulations (AWWA 2000). These changes in treatment practices could be as simple as controlling pH or could be a complete overhaul of existing water treatment facilities. This chapter describes some common treatment technologies and their effectiveness in mitigating regulated contaminants in raw water sources.

DEFINITION AND NEED FOR BEST AVAILABLE TECHNOLOGIES (BATs)

The National Primary Drinking Water Regulations (NPDWRs) promulgated in recent years are complex and in some cases impose conflicting objectives on a water system (Pontius 2003a). Integrated assessment for compliance is important, especially if capital improvements will be needed, because the process to evaluate regulatory effects, conduct pilot studies, design needed facilities, develop financing and complete construction typically takes a few years.

The selection of a particular technology for drinking water treatment is governed largely by the PWSs consideration of a number of constraints in addition to the requirement to provide safe drinking water. Sometimes the best technology for a particular source-water or regulated contaminant may not be feasible due to other considerations. These considerations may include state and national environmental and permitting restrictions on the plant-siting, limitations on withdrawals from a desirable source, restrictions on disposal of process wastes, logistic requirements of monitoring, operation and maintenance, future capacity development and cost of construction (Pontius 2003b).

The USEPA provides some of this information to small systems under the 1996 SDWA Amendments authorization through Small Systems Compliance Technology Lists. Under this authorization, the USEPA is required to provide technical and economic guidance to PWSs for compliance with the MCL for each regulated contaminant. To achieve this, the USEPA identifies certain treatment technologies that have demonstrated efficient and economic removal of regulated contaminants. These economically achievable technologies for regulated contaminants are known as 'best available technologies' (BATs).

The BATs are known to work for particular contaminants in some source waters if designed and operated properly. However, due to the diverse source-waters and their treatment requirements, systems have to use their own judgment in the use of the BATs. Though the BATs identified by the USEPA may not always be the best solution to specific water supplies, they provide a useful baseline for the acceptability of available and future treatment technologies. Use of BATs provides (in some cases) a body of evidence that adequate contaminant mitigation measures are being carried out by the PWS. However, the ultimate criterion for the acceptance of any treatment process is a demonstrated removal of a specific contaminant below the MCL. For water treatment technologies that are not identified by the USEPA as BAT for a specific contaminant, utilities must demonstrate the efficacy of the process through pilot studies (Pontius 2003a).

CONTAMINANT SPECIFIC TREATMENT TECHNOLOGIES

The choice of treatment techniques is usually determined by a combination of factors, including the type of source waters (ground, surface or saline water) – which determines the raw water quality (levels and types of contaminants) – and the quantity and quality of residuals generated.

The common practice is to employ a number of unit treatments in a train in order to achieve removal of a wide range of contaminants. These unit treatments within the treatment train are located in a specific sequence to gain maximum removal efficiencies. To achieve this, the source water is characterized by measuring the level of all regulated contaminants in the source water. The treatment train is then selected by incorporating unit processes that can mitigate the maximum number of contaminants in the source water. Unit processes thus selected have to be compatible and minimize the need for the conditioning of the feed waters. For instance, the integrated membrane systems for surface water treatment incorporate coagulation of source water followed by settling and nanofiltration. The coagulation process removes the organic content of the surface water while the nanofiltration process reduces the ionic content of the water and provides a barrier for pathogens.

The selection of a particular train of unit processes is complicated as even a small number of unit processes can be arranged in numerous combinations (trains). The decision to select a particular train also has to be tempered by the fact that depending upon their location within the treatment train, unit processes can be synergistic or antagonistic for the removal of specific contaminants. For instance, the practice of prechlorination of feed water for slow sand filtration may be helpful in mitigating uncontrolled growth of the biological layer. However, sustained chlorination at a high dosage may damage the biological layer and thereby affect the removal of pathogens adversely. In this instance the same practice can either help or hinder depending upon the level of chemical dosage (Imran 1997).

The evaluation of all different combinations of unit processes and their related water quality impacts is neither intended nor attempted in this project. Instead, different unit processes are evaluated for their simultaneous impact on the removal of different contaminants. The water quality impacts for the unit processes comprising a train can be collated to give an estimate of the overall water quality impact due to the treatment train. This assumes that the water quality impact of a particular treatment train can be obtained by the aggregation of the individual impacts of the unit processes that form the treatment train.

Control of Organic Contaminants

Organic contaminants are classified as natural organic matter (NOMs), synthetic organic contaminants (SOCs) and volatile organic contaminants (VOCs). Where, SOCs and VOCs are regulated for their adverse health effects, NOMs are indirectly regulated through the D/DBP Rules and taste and odor (T&O) guidelines. Other classes of organic contaminants like assimilable organic carbon (AOC) and biodegradable dissolved organic carbon (BDOC) have received recognition as promoters of growth of biofilm and bulk water microorganisms. Though their exact interaction with microbial populations in a distribution system is yet to be conclusively established, it is recognized that they play an integral role in the microbial dynamics within the distribution system.

There are two main technologies that have demonstrated high removal efficiencies of all the major classes of organic contaminants: activated carbon adsorption and packed tower aeration (Pontius 1991). Other technologies like membrane treatment (reverse osmosis, nanofiltration and ultrafiltration) have also demonstrated high organics' removal under field and laboratory conditions. However, membrane technologies are primarily used for the removal of ionic contaminants from source waters and are generally not designed for the removal of organics due to increased fouling under high organic loading.

Processes based on activated carbon adsorption are used extensively due to their ease of use and the ability to remove a wide range of organic contaminants. Activated carbon is similar to charcoal in composition, but is surficially altered to enhance its sorption properties. The base carbon material (typically bituminous coat, lignite, petroleum coke, or bone char) is carbonized through heating it in the absence of air and then activated by oxidation at high temperature (200 to $1000^{o}C$). The activation process results in a highly porous, high surface area per unit mass material with a large sorption capacity for many water impurities, including SOCs as well as T&O causing compounds (Snoeyink and Summers 1999). Modifications of the activated carbon technology include granular activated carbon (GAC), powdered activated carbon (PAC) and biological activated carbon (BAC). Removal efficiencies for many organic contaminants are excellent under ideal circumstances, but can be significantly affected by water quality parameters, like dissolved natural organic matter levels, pH, and temperature.

Control of Inorganic Contaminants

A number of technologies are available for the treatment of inorganic contaminants in source waters. Traditionally, utilities have successfully complied with the MCLs for inorganic contaminants in source waters. Treatment for inorganic contaminants usually involves the chemical exchange of an undesirable contaminant of the source water with a benign one.

For instance, ion exchange (IX) is a unit process used to selectively remove undesired ions such as calcium (Ca^{2+}), magnesium (Mg^{2+}), iron (Fe^{2+}), manganese (Mn^{2+}), ammonium (NH_4^+), nitrate (NO_3^-), and fluoride (F^-) (Clifford 1999; Clifford, Ghurye, and Tripp 2003; Ghurye, Clifford, and Tripp 1999). Ion exchange medium consists of a solid phase and a mobile ion attached to an immobile functional group. The mobile ions in the resin are exchanged with ions in the water. The spent resin is then regenerated and reused. Ion exchange resins are anionic or cationic. Cationic resins exchange positive ions and anionic resins exchange negative ions. Anion exchange resins selectively remove anionic species such as NO_3^- and F^-. Anion exchange resins are often regenerated with sodium hydroxide ($NaOH$) or sodium chloride ($NaCl$) solutions, which replace the anions removed from the water with hydroxide (OH^-) or chloride (Cl^-) ions, respectively. Cation exchange resins are used to remove undesired cations from water and exchange them for hydronium (H^+), sodium (Na^+) or potassium (K^+).

Membrane technology, such as nanofiltration, reverse osmosis or electrodialysis/ electrodialysis reversal is capable of removing inorganic contaminants by providing a physical/ electro-chemical barrier between the source and treated permeate. These technologies, especially reverse osmosis, are capable of removing more than 90% of dissolved inorganic contaminants from source waters. Membrane treatment is non-selective, since it removes all inorganic constituents (including contaminants) from the source water. Membrane treatment produces aggressive water that needs to be stabilized prior to distribution.

Control of Microbial Contaminants

Researchers in the US have made a compelling argument for maintaining disinfectant residuals for North American water supplies (Haas 1999a; LeChevallier 1999). In addition to the primary role of disinfection at the treatment stage, the secondary role as a residual is important to avoid regrowth or intrusion problems. Microbial contaminants differ from other organic and inorganic contaminants in that their levels, occurrence and sources cannot be precisely predicted in time for proactive mitigation measures. Under the SWTRs and the upcoming GWR, utilities are expected to estimate the baseline level of pathogens in their source water and provide removal measures based on the source water levels. Additional protection after treatment is provided, by requiring covered reservoirs, minimum disinfectant residuals and monitoring for pathogens in the distribution system (under the TCR).

Filtration is the most common treatment of choice for the removal of microorganisms from source waters. However, filtration is often practiced in conjunction with other unit processes such as coagulation, lime softening and disinfection. Filtration is almost always followed by disinfection due to the increased potency of disinfectants in filtered waters.

Disinfection, especially chlorination, has been the most common treatment in many water systems to provide primary and secondary disinfection. However, concern over chlorinated DBPs has prompted the need for alternate disinfectants. Chloramination, ozonation, chlorine dioxide, UV irradiation and advanced ozone/peroxide processes have been gaining ground as the alternate treatments. However, some (chloramination) lack potency, some (ozonation) do not provide residual protection and most are complex in application compared to chlorination.

The choice of the disinfectant is also determined by the source water quality and the potential for forming byproducts that might impact consumer health (USEPA 2001a).

Control of Contaminants Generated in Distribution Systems

Contaminants generated in the distribution system are due to the interaction among bulk water constituents (residual chlorine and organic matter react to form THMs and HAAs), interaction with pipe surfaces (corrosion byproducts, biofilm slough-off) and hydraulic effects.

The internal surfaces of distribution pipes can act as a sink and/ or source of chemical and biological contaminants (Fisher et al. 2000; Lytle, Sorg, and Frietch 2004; Valentine and Stearns 1994). Changes in water quality can trigger the release of these contaminants to an extent where the water quality becomes unacceptable for use even though the water quality at the treatment plant was acceptable (Imran et al. 2005b).

All physical, chemical and microbiological processes that take place in the distribution system are related in some way to the quality of the water being distributed. Over a period of historical usage any source-water will establish an "apparent" equilibrium of physico-chemical and biological reactions in the distribution system. This apparent equilibrium makes it possible to predict and maintain a specific quality for the water being distributed. Recent trends in source-water diversification, regulatory restrictions and demand dynamics may result in distribution systems receiving variable water quality. These water quality changes may be either gradual or sudden depending upon the operational conditions. Any change in the finished water quality will disrupt the existing equilibrium resulting in a transitional state until a new equilibrium is achieved. This new equilibrium may either improve or deteriorate the water quality at the consumers tap (Taylor et al. 2005). Distribution systems that have maintained an historic single source for drinking water supplies can have problems maintaining the distributed water quality when alternate sources are used (Lovins, III et al. 2005; Marangou and Savvides 2001; Taylor et al. 2005).

An AwwaRF study on balancing multiple water quality objectives identifies the different constraints governing feasible treatment alternatives as water quality, regulatory, economic and operational (Daniel 1998). A similar framework can be constructed for the regulatory restrictions on use of treatment technologies and the sometimes conflicting water quality requirements to comply with different water quality regulations as shown in Figure 3.1.

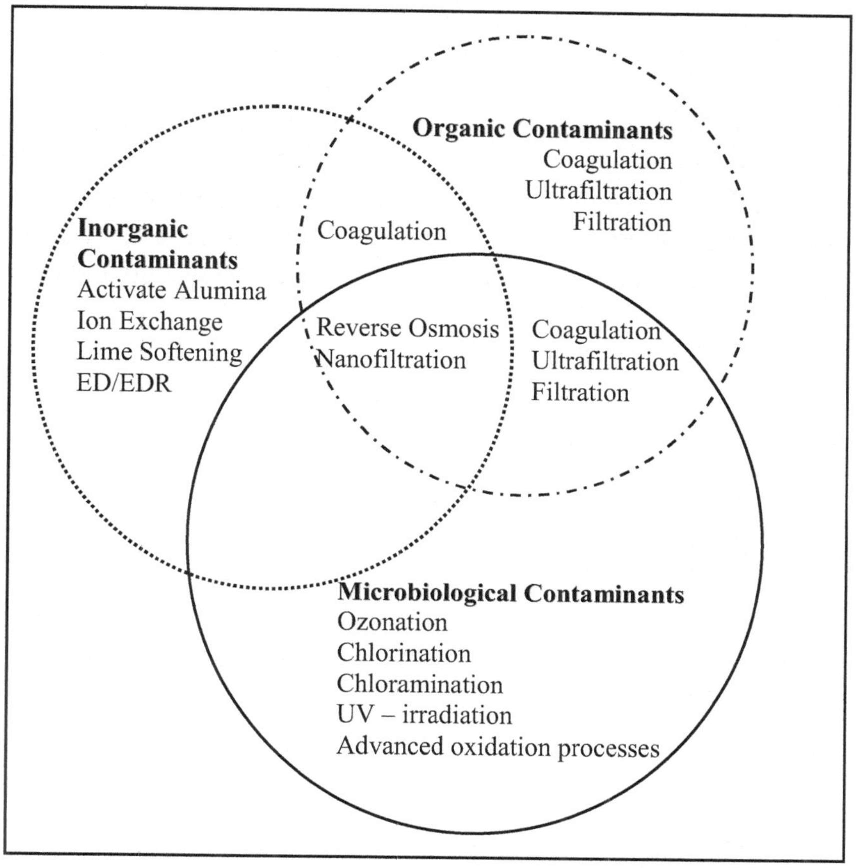

Figure 3.1 Selection of contaminant specific treatment technologies

RELATIONAL MATRIX (UNIT TREATMENT VS. CONTAMINANT REMOVALS)

A relational matrix of unit treatment-regulation (**A**) is developed, which relates the effectiveness of common unit treatments to compliance of drinking water regulations. The element α_{ij} is a measure of the effectiveness of unit treatment '*j*' (UT_j) for the compliance of drinking water regulation '*i*' ($\mathrm{Re}g_i$). The effectiveness of different unit technologies in achieving desired removals of the four classes of contaminants is given in Table 3.1. The effectiveness estimate is a number between +1 and –1. A value of +1 indicates the most effective technology to remove the selected contaminant, while –1 indicates the technology that might have an adverse effect in removal of a selected contaminant. A value of 0 is given for the technology that has no known effect on the removal of selected contaminant.

The treatment technologies indicated in Table 3.1 are generic with respect to the contaminant targeted and the mechanisms of contaminant removal. For instance, activated carbon may be used as powdered activated carbon (PAC) or granular activated carbon (GAC). Though the mechanism of removal of the targeted contaminant is similar in both, the method of application and design parameters are different. The removal levels attributed to these generic technologies are the maximum removals possible under ideal design and operational conditions.

Table 3.1
Effectiveness of selected treatments in removing drinking water contaminants (A)

Contaminant Categories	Air Stripping[a]	Activated Alumina[b]	Activated Carbon[c]	Chlorination[d]	Chloramine	Ozonation	Advanced Oxidation	Coagulation[e]	Ion Exchange[f]	Precipitation /Softening[g]	Filtration	Membranes[h]
Inorganics												
Antimony								0.8				0.9
Arsenic[i]		1.0[j]		[k]	[l]	[m]		0.8[n]	0.8[o]	0.8[p]	[q]	0.9[r]
Barium									0.8	0.8		0.9
Beryllium, Chromium		0.8						0.8	0.8	0.8		0.9
Cyanide[s]				0.8					0.9			0.9
Fluoride		0.8[t]										0.8
Lead and Copper[u]	[v]											
Mercury (inorganic)			0.8					0.8		0.8		0.8
Nitrate + Nitrite									0.8			0.8
Radon	0.8[w]											
Selenium		0.8[x]						0.8	0.8	0.8		0.8
DBP Formation Potential												
Bromate			0.4[y]			-1.0	-1.0					
TTHMs			1.0[z]	-1.0	-0.2[aa]							0.8[bb]
HAA5			1.0	-1.0	-0.2							0.8
Microorganisms[cc]												
Total coliforms			[dd]	1.0[ee]	0.8	1.0	1.0	0.6				0.8[ff]
HPC bacteria				1.0	0.8	1.0	1.0	0.6			0.5	0.8
Giardia				0.5[gg]	0.3	0.7	1.0	0.6				0.8
Legionella				0.8	0.8	1.0	1.0	0.6				0.8
Viruses				0.8[hh]	0.5	0.9	1.0	0.6				0.8
Cryptosporidium				0.8	0.5	1.0	1.0	0.6				0.8
Turbidity				0	0	0		0.8			0.8	
Organic Contaminants												
Volatile organics	1.0[ii]		0.8[jj]			0.5						0.5
Synthetic organics	0.8		1.0[kk]			0.5		0.8[ll]				0.8[mm]
Dissolved organics	0.5		1.0[nn]			0.5		0.8				0.8[oo]
Radionuclides												
Radium 226 + 228									0.8[pp]	0.8		0.8
Gross alpha, total beta and photon particle activity										0.8		0.8

Effectiveness factor criterion is as follows:

+1.0 Most Effective Technology — 0 No Effect — -1.0 Adversely Effective Technology

Blank No data

[a] Air stripping is effective in removing most volatile organic contaminants (VOCs), radon and dissolved gases. However, off-gas treatment is required in most cases. Hand, Hokanson, and Crittenden 1999

[b] Clifford 1999

[c] Snoeyink and Summers 1999

[d] Haas 1999b

[e] Letterman, Amritrajah, and O'Melia 1999

[f] Clifford 1999

[g] Benefield and Morgan 1999

[h] Membrane filtration is most effective in removing inorganic contaminants. Organic contaminants can cause increased fouling and loss of efficiency. Taylor and Wiesner 1999

[i] Thirunavukkarasu et al. 2005; USEPA 2003

[j] Gurian and Small 2002; Kunzru and Chaudhuri 2005;Sutherland et al. 2002; USEPA 2003; Wang et al. 2002

[k] Preoxidation may be required to convert As(III) to As(V) prior to further treatment. Edwards et al. 1998; Ghurye and Clifford 2004

[l] Ghurye and Clifford 2004

[m] Ghurye and Clifford 2004

[n] Chen et al. 2002; Chwirka et al. 2004; Ghurye, Clifford, and Tripp 2004; Gregor 2001; Han et al. 2002; USEPA 2003

[o] Ghurye, Clifford, and Tripp 2004; Kim et al. 2003; USEPA 2003; Wang et al. 2002

[p] Chen et al. 2002;Edwards et al. 1998; USEPA 2003

[q] Filtration for Arsenic removal has to be preceded by coagulation (softening) process. Chwirka et al. 2004; Ghurye, Clifford, and Tripp 2004; Han et al. 2003; USEPA 2003

[r] Ning 2002; Oakes 2005; USEPA 2003; Van der Bruggen and Vandecasteele 2003

[s] USEPA 1998a

[t] Ghorai and Pant 2004; Ghorai and Pant 2005

[u] Lead and Copper are regulated by treatment technique under the Lead and Copper Rule

[v] Edwards, Schock, and Meyer 1996; Jacobs, Reiber, and Edwards 1998; Lytle et al. 1998

[w] Hand, Hokanson, and Crittenden 1999

[x] Pyrzynska, Drzewicz, and Trojanowicz 1998

[y] Kirisits et al. 2001;Kirisits et al. 2002

[z] Bond and DiGiano 2004; USEPA 1999a

[aa] Diehl et al. 2000

[bb] Clark et al. 1998; Magara et al. 1996; Matilainen et al. 2004

[cc] USEPA 1999a; USEPA 2001a

[dd] Lau et al. 2005

[ee] USEPA 1999b; USEPA 2001a

[ff] Coliforms are regulated in distribution systems through the Total Coliform Rule. Technologies other than residual disinfectant may effect coliform growth by controlling nutrients or increasing disinfectant efficacy. USEPA 1998b

[gg] Haas et al. 1998; Smith et al. 1995

[hh] Haas 1999b; Springthorpe et al. 2001; Thurston-Enriquez et al. 2003

[ii] Hand, Hokanson, and Crittenden 1999; Narbaitz et al. 2002

[jj] Campos et al. 2000

[kk] Ayele, Levavasseur, and Mazet 1996; Griffini et al. 1999; Lee et al. 2003; Sorrell, Brass, and Reding 1980

[ll] Acero and Gunten 2001; Acero et al. 2001;Rosenfeldt and Linden 2004

[mm] Acero, Rodriguez, and Meriluoto 2005; Chen et al. 2004

[nn] Campos et al. 2000;Karanfil and Kilduff 1999; Paune et al. 1998

[oo] Alborzfar, Jonsson, and Gron 1998; Anselme et al. 1993; Choo, Park, and Choi 2004; Leiknes, Odegaard, and Myklebust 2004; Taylor and Wiesner 1999

[pp] Sorg and Logsdon 1980

SECONDARY WATER QUALITY IMPACTS OF TREATMENT TECHNOLOGIES

The treatment process or modifications to the treatment process impact the distribution system (Lytle et al. 1998). The primary objective of any treatment technology is to achieve acceptable removals of the targeted contaminants. However, the use of pre-treatment, treatment and post-treatment practices cause changes in the ionic content of the water and can have adverse impacts on distribution infrastructure. For instance, an anion exchange resin exchanges bivalent metallic ions with monovalent sodium ion. In this instance, ion exchange treatment would be responsible for removing a known inhibitor of corrosion in iron pipes (calcium and magnesium hardness) with sodium – resulting in a possible increase in the corrosion rate of unlined iron surfaces.

No single unit technology is effective in removing all the drinking water contaminants to a safe drinking level. Treatment technologies that are BAT for a selected contaminant may not be effective in removing other co-occurring contaminants. Therefore, to achieve simultaneous removal of a wide range of contaminants utilities use unit treatment technologies placed in a sequential order to remove the maximum number of contaminants.

Another complicating factor is the use of different chemicals for similar objectives. Identifying all the secondary changes in water quality is an impossible task due to the numerous treatment practices. For instance, coagulation may be achieved by using alum, ferric chloride or ferric sulfate. Alum and ferric sulfate will increase the sulfate in the finished water relative to the source water, while ferric chloride will increase the chloride content of the water. Both sulfate and chloride increase the corrosion of metallic pipes, albeit to a different degree. Coagulation is also accompanied by a consumption of total alkalinity in the source water. Decreased alkalinity is detrimental to iron components in the distribution system, but at the same time may be beneficial in reducing copper corrosion.

Secondary changes induced by the treatment technologies are also dose specific. For instance, depending on the source water quality, different concentration of treatment chemicals need to be added and thereby result in different concentrations of secondary ions in the finished water.

A relational matrix (**B**) for unit technology Vs water quality parameters is developed which relates the contribution of unit technology UT_j to increase or decrease in selected water quality parameters WQP_k. The water quality parameters selected are those known or suspected in the deterioration of distribution infrastructure.

Table 3.2 gives an estimate of the magnitude of the secondary water quality changes that occur due to the use of common treatment technologies.

Table 3.2
Secondary water quality impact of treatment techniques and practices (B)

	pH	Calcium	Alkalinity	Sulfate	Chloride	Dissolved Oxygen	o-, p-phosphate	Silicates	Total Dissolved Solids	Ammonia	Chlorine	Chloramine	Natural Organic Matter	Biofilms
Air Stripping[a]	0.3[b]	-0.2				0.8			-0.2	-0.5			-0.3[c]	
Activated Alumina[d]	-0.5		-0.5											
Activated Carbon[e]													-1.0	
Chlorination[f]														
Cl$_2$ gas	-0.3		-0.2		0.5				0.1	-1.0	1.0			-0.7
NaOCl	0.3		0.2		0.5				0.1	-1.0	1.0			-0.7
Chloramination	0				0.5				0.1	0.3		1.0		-0.5
Ozonation						0.8				0[g]			-0.5	0.5
Advanced Oxidation						0.8								
Coagulation[h]														
FeSO$_4$	0.8		-0.5	0.8									-0.8	
Alum	-0.3		-0.5	0.8									-0.8	
FeCl$_3$	0.8		-0.5		0.8								-0.8	
Ion Exchange[i]		-0.8							-0.5					
Precipitation/Softening[j]	0.8	-0.5	-0.5						0				-0.4	
Filtration[k]													-0.6	
Membranes[l]	-0.5	-0.8	-0.8	-0.7	-0.5				-0.8				-0.9	
Internal Corrosion Control[m]														
pH	0.8		0.2											
o-, p-phosphates							0.8							
Silicates								0.8						

Effectiveness Factor criteria is as follows:

+1.0 Definite Increase — 0 No Effect — -1.0 Definite Decrease

Blank — No data

[a] Hand, Hokanson, and Crittenden 1999

[b] Lytle et al. 1998;Schock et al. 2002

[c] Narbaitz et al. 2002

[d] Clifford 1999

[e] Snoeyink and Summers 1999

[f] Haas 1999b

[g] Haas 1999b

[h] Letterman, Amritrajah, and O'Melia 1999

[i] Clifford 1999

[j] Benefield and Morgan 1999

[k] Cleasby and Logsdon 1999

[l] Taylor and Wiesner 1999

[m] Schock 1999; AwwaRF and DVGW 1996

CHAPTER 4
DISTRIBUTION INFRASTRUCTURE INTEGRITY

The deterioration of drinking water distribution infrastructure is among the main causes of the loss of quality and quantity of water at the consumers tap. However, since a major portion of the distribution infrastructure is underground, its deterioration does not present the same visual urgency as other visible infrastructure. Since deterioration of the distribution infrastructure adversely impacts the water quality, PWSs cannot justify the costs and efforts of treating water to potable levels and then transmitting them through deteriorating distribution systems. Some of the most critical water quality issues related to water distribution infrastructure deterioration were highlighted in a series of white papers produced by the USEPA, AWWA (American Water Works Association) and AWWSC (American Water Works Service Company). These white papers underline the importance of pro-active management of distribution systems to maintain water quality at an acceptable level (Scharfenaker 2002).

Water distribution infrastructure is a complex system that includes all components necessary to transport potable water from the treatment plant to the consumer. These components can be of various materials, placed in different configuration and having different functions. This diversity of components and their functions makes it difficult to define the extent of deterioration in a definitive manner. For instance, different priorities may be assigned to a cast iron pipe or a pump casing even though both may exhibit a similar extent of deterioration. However, in this research the impact of water quality on distribution infrastructure is mainly related to its interaction with the component surface. Where the infrastructure is comprised of components of different materials, a holistic determination of water quality impacts is purely qualitative in nature.

This chapter summarizes a survey of research related to the impact of water quality on common distribution infrastructure materials. Though an effort was made to separate the water quality impacts on material from the material impacts on water quality, the readers should realize that this duality is inherent in the distribution system. Therefore, where independent information for water quality impacts on distribution materials is not available, water quality deterioration has been taken as a surrogate measure for deterioration of distribution infrastructure.

COMPROMISED DISTRIBUTION SYSTEM INTEGRITY – HOW BIG IS THE PROBLEM?

Any distribution component will deteriorate due to internal and external corrosion, wear and tear, fatigue and 'problems of age'. Deterioration rate and extent is a function of stresses to which the pipe is exposed and the duration of exposure. Though chronological age of a pipe alone cannot determine the deterioration, it can be an indicator of the duration of exposure. The term 'age' of water mains in general is used as a surrogate for 'deterioration' rather than chronological age. The failure of water distribution infrastructure can cause an array of economic costs including direct damage to transmission and distribution infrastructure, adjacent property and business, traffic delays and inconvenience to public (Cromwell III et al. 2002).

Table 4.1
Percent distribution of pipe materials used in Europe and North America (Rajani 2006a)

Europe	**CI**	**AC**	**Plastic**	**Steel**	**Concrete**		
UK	81	10	6	2	1		
Switzerland	80	10	3	7			
Sweden	61	28	3	3	5		
Luxemburg	58	6	12	24			
Germany	55	9	30	5	1		
Ireland	53	20	25	1	1		
Denmark	50	10	35	1	4		
Norway	44	15	38	2	1		
Belgium	40	33	10	13	4		
Austria	33	30	32	5			
Spain	30	50	15	4	1		
Finland	17	3	76	3	1		
Netherlands	16	40	40	3	1		
North America	**CI**	**AC**	**PVC**	**Steel**	**Concrete**	**DI**	**Other**
Northeast	61	25	4	3	2	2	3
Southeast	41	23	9	2	2	19	4
Midwest	57	22	5	1	2	12	1
South central	44	14	18	2	4	13	5
West	26	6	43	18	2	2	3
Northwest	41	26	14	12	1	5	2
Total US	46	20	14	4	2	11	3
Canada	42	24	15	1	4	12	2

Table 4.1 summarizes the data of percent representation of various pipe materials in North America and in different countries of Europe. The cast and ductile iron pipes constitute a major portion of the distribution network material. However, utilities increasingly prefer polymeric materials to prevent aesthetic water quality failures.

According to an estimate of the United States water main breaks data reported by Kirmeyer, Richards, and Smith (1994), there are approximately 0.17 water main breaks per km of pipe per year for all types of piping. The corresponding pipe break estimate for Canada is 0.20 breaks per km of pipe per year (Table 4.2). A further breakdown of the pipe breaks data for Canada shows that the highest number of breaks occurred for cast iron pipes followed by ductile iron, PVC and AC pipes. However, if percentage of different pipe materials (Table 4.1) is considered, it appears that ductile iron pipes have the highest breakage rate followed by PVC, AC and CI. The pipe-break problem can lead to the contamination of the distribution system that may occur prior to, during and after the repair and rehabilitation of these breaks (Deb et al. 2000).

Another manifestation of distribution infrastructure deterioration is the prevalence of leaks due to failure of infrastructure components such as tanks, pipes and fittings. Though leaks do not present a catastrophic breakdown compared to breakage failures, they are expensive for a variety of reasons, including the loss of water and treatment chemicals, the increased risk of water quality deterioration (in case of contaminant intrusion), unnecessary capital expansion and increased energy expenditure required to feed the leak (Colombo and Karney 2002). Contamination may occur through leaks during short-term pressure transients in the distribution system. When internal pressure in the water main drops, contaminated water (or soil) near the water main may flow in through leakage points (Boyd et al. 2004a; Boyd et al. 2004b; Karim, Abbaszadegan, and LeChevallier 2003). Contaminants may also intrude through deficiencies in the distribution system appurtenances like joint seals and flooded vaults or meter boxes (Sadiq, Kleiner, and Rajani 2006).

The water quality impacts on infrastructure surfaces in contact with the finished water are by no means the only (or even the most significant) process that leads to the overall deterioration of the infrastructure components. Other processes including, structural loading, external corrosion, inadequate operation and maintenance and human errors are significant causes of infrastructure failure. However, the water quality induced deterioration may exacerbate the condition of the pipe and make it susceptible to other failure mechanisms.

Water quality deterioration in distribution system is inextricably linked to the condition of the surfaces in contact with the finished water. Where water quality problems exist, these can be directly attributed to the extent of deterioration of the distribution infrastructure. Though repair and rehabilitation measures can mitigate the water quality problem in the short-term, an assessment based on the water quality interaction with distribution infrastructure is essential to realize a long-term solution to infrastructure deterioration.

Table 4.2
Comparison of breakage rate data for various pipe materials collected from Canadian municipalities (# of breaks/km/yr) (Rajani 2006b)

	Materials			
Statistics	CI	DI	PVC	AC
Mean	0.30	0.25	0.14	0.97
Standard deviation	0.12	0.19	0.96	0.11

Many complex interactions occur simultaneously between the finished water and the distribution surfaces in contact that contribute to this deterioration. For instance,

- Conduit surfaces in contact with water corrode and the corrosion byproducts are transported and deposited near dead-ends and low-flow regions.
- Biofilms on the conduit surface contribute to microbiologically induced corrosion (MIC), and may slough-off causing microbiological proliferation.
- Contaminants in the water may degrade or accumulate over time due to interaction with pipe surfaces and/or bulk reactions.

DETERIORATION PROCESSES

Distribution components could be of different materials including cast iron, ductile iron, steel, copper, lead, galvanized steel, reinforced concrete, asbestos cement, thermoplastics including polyvinyl chloride (PVC), chlorinated polyvinyl chloride (CPVC), high density polyethylene (HDPE), polybutadiene and composites such as glass fibre-reinforced plastic (GFRP). In addition to the pipes and mains, there are a number of ancillary materials like coatings, gaskets, o-rings, fittings, valves and solders. Additional variability is introduced by the presence of man-made (cement, epoxy, polymeric, or calcite lining) as well as naturally occurring films (corrosion byproducts and biofilms) on the interior of the pipe (Broo, Berghult, and Hedberg 2001b; Kirmeyer, Richards, and Smith 1994; Tomboulian et al. 2004).

The problem of identifying the impacts on a complex system can be simplified by evaluating the most common repeating sub-system or system component. For the purpose of evaluating the impact of water quality on the integrity of distribution infrastructure, the most common denominator is the material of construction. Though the distribution infrastructure is composed of different components fulfilling different functions, they are constructed using an exhaustive list of materials. The choice of materials in drinking water applications is limited by the health and aesthetic concerns as well as economic considerations of procurement, installation, operation and maintenance (Broo, Berghult, and Hedberg 2001b).

Depending on the material make-up of infrastructure components, the distribution infrastructure can be divided into three broad groups - metallic, polymeric and/or cement-based. The distinction between these three is based on the dominant deterioration processes; (physico-chemical based) internal corrosion, microbiologically induced corrosion (MIC) and leaching of material. Though a specific material may have a predominant mechanism of deterioration, all three processes play a significant role towards its overall deterioration (Table 4.3).

Physico-chemical based internal corrosion (from this point onwards referred to as 'corrosion') is defined as 'metallurgy in reverse', where a purified metal or its alloy interacts with the environment to return to its original more stable state. Three conditions are required in order for corrosion to proceed; a metallic surface that will corrode, an oxidant that will oxidize (corrode) the metal to a more stable state and lastly, a medium that will transport the oxidant to the metal and facilitate further corrosion by moving the corrosion byproducts away from the corrosion site. All these components are present in a typical water distribution system. A water distribution system may corrode both internally and externally. However, the external corrosion is not directly related to the water quality issues and therefore is not discussed here.

Table 4.3
Significance of impact of water quality deterioration mechanisms on distribution materials

		Internal Corrosion	**Microbiologically Induced Corrosion**	**Leaching**
Water Distribution Infrastructure Material	**Metallic Components** (Iron, copper and lead)	Major	Unknown	Minor
	Polymeric Components (PVC, PE and PAH)	None	Unknown	Major
	Cement-based Components (AC and CC)	Major	Unknown	Major

AC – Asbestos cement, CC – Concrete, PE – Polyethylene, PAH – Polyaromatic hydrocarbons (bituminous or coal-tar), PVC – Polyvinylchloride

Microbiologically induced corrosion is different from other deterioration processes because it is caused by the biological activities of microorganisms. While some researchers suspect that biofilms (colonies of native microorganisms on the water/ metal interface) inhibit deterioration by providing a protective coating, others have reported biofilm as exacerbating deterioration. The presence and structure of these biofilms is related to the hydraulic conditions, nutrient availability, type and concentration of residual disinfectant and the roughness of the pipe surface. Many researchers have reported that pipe material seems to have a significant effect on microbial inactivation by different disinfectants (Gagnon et al. 2005).

Leaching is typically defined as the release of material to water without involving corrosion processes. It can take the form of dissolution of the metal-bearing corrosion scales, monomers from plastics, or calcium from the cement-matrix.

Deterioration of Metallic Components

Deterioration of metallic components, almost always, occurs due to corrosion. Corrosion of metallic components can cause three distinct but interconnected problems. First the metal is lost through oxidation to soluble metallic species or metallic corrosion scales leading to loss of structural strength and subsequent leaks and breaks (Boulos et al. 2005,Colombo and Karney 2002, Lin, Ellaway, and Adrien 2001). Second, the scale can accumulate as large tubercles that increase head loss and decrease hydraulic capacity (Larson and Skold 1957). Finally, the release of soluble or particulate metallic corrosion byproducts to the water bulk decreases its acceptability for potable use (Edwards, Jacobs, and Taylor 2000; Imran et al. 2005b; McNeill and Edwards 2001).

Biofilms on metal surfaces produce an environment at the biofilm/metal interface that is radically different from that of the bulk medium in terms of pH, dissolved oxygen, organic and inorganic species (Little, Wagner, and Mansfield 1992; Little and Ray 2002). Microorganisms can accelerate rates of partial reactions in the corrosion process or alter corrosion mechanisms. The physical presence of biofilm causes variable conditions on the metal surface resulting in differential aeration and the formation of consequent anodic and cathodic regions that can accelerate corrosion. Uniform biofilms can decrease corrosion by providing protective film, whereas the products of biological activity can either enhance or reduce corrosion (Gagnon et al. 2005).

Surface film and scale are naturally occurring deposits or growths that occur on the interior pipe surfaces as a result of corrosion, biological growth or residue deposition. The composition and structure of the surface film depends on the type and composition of the pipe and the aqueous phase. The formation of surface film is governed by several factors that include pH, alkalinity, buffer intensity, natural organic matter (NOM) and dissolved oxygen (DO). Sarin et al. (2001) relate the effect of water quality parameters on metal release from corrosion scale to the modification of the properties of the corrosion scale; namely porosity and solubility of scale phases.

Iron

For many years, starting in the late 1800s iron has been the material of choice for the construction of distribution piping. Currently, nearly 70% of the distribution mains in North America are iron (Table 4.2). Some of the most persistent water quality problems are due to the corrosion of unlined cast iron pipes and lined iron pipes that have lost their interior linings. Manufacturing practices and exposure to corrosive environment are the major contributors to the deterioration in cast iron components. Besner et al. (2001) report that the maximum number of pipe breaks in a study area in Montreal, Quebec, occurred in pipes that were installed between 1963 and 1974. The lowest pipe breaks occurred in pipes that were installed prior to 1900. The authors attributed this to the difference in material used and workmanship.

Though conceptually related, the effect of internal corrosion on iron release to water is tenuous and depends on the finished water quality (Sarin et al. 2001; Sarin et al. 2004b). For instance in some iron pipes, the metal (corrosion byproducts) release is proportional to corrosion rates e.g., for low pH, low buffer capacity waters. In others the relationship is not so clear, e.g., for high pH, high dissolved oxygen. In the former case low pH causes metallic solubility and is seen as iron release. While in the latter case (typical of most water systems), though higher corrosion rates may exist due to high levels of dissolved oxygen the metal released may precipitate within the corrosion scales due to high pH (Sarin et al. 2004a). The entrapped corrosion byproducts can protect the iron surface from further corrosion by providing a denser and less porous surface scale (Sarin et al. 2003). The surface scales also provide a potential reservoir of corrosion byproduct that can release under system disturbances (Imran et al. 2005b; Lehtola et al. 2004b; Sarin et al. 2004b).

The problem of colored water release is often confused with the corrosion of metallic pipes. In old metal pipes colored water problems are mainly due to dissolution and scour of internal corrosion scales. This disruption may be due to changes in water chemistry and/ or flow patterns within the distribution systems. Red water problems occur predominantly due to the release of corrosion byproducts from old cast iron and galvanized iron pipes (Imran et al. 2005b). The iron released may be due to either direct corrosion of the pipe material or due to the dissolution and/or scour of corrosion scales in the interior. The release of iron is mainly in the form of particulates and flakey material from the corrosion scales formed on these pipes (Baylis 1926; Lin, Ellaway, and Adrien 2001; Sarin et al. 2001; Sarin et al. 2004b). Similar release mechanisms lead to blue water release from copper pipes (Edwards, Jacobs, and Taylor 2000).

The problems associated with colored water conditions are mainly aesthetic. However, a more serious consequence could be the loss of disinfectant residual and subsequent biofilm formation in the distribution systems (Frateur et al. 1999; Hallam et al. 2002; Vikesland et al. 2000; Vikesland and Valentine 2002a). Corrosion byproducts attached to pipe surfaces or accumulated as sediments in the distribution systems can shield microorganisms from disinfectants (Butterfield et al. 2002; Holden et al. 1995; LeChevallier 1999; Lehtola et al. 2005b). Release of corrosion byproducts in the form of turbidity may provide safe havens for pathogens (Morin et al. 1999; Torvinen et al. 2004).

The mineral composition of the source water usually determines whether adverse conditions (that will lead to corrosion and subsequent deterioration) exist within the distribution systems (Imran et al. 2005a). Biological conditions (that control MIC) are also inextricably linked to the mineral content of the source water (Bade, Manz, and Szewzyk 2000; Carter et al. 2000; Melo and Vieira 1999). Increase in neutral ions like sulfates and chlorides cause a corresponding increase in corrosion. This phenomenon is hypothesized to be due to the disruption of protective corrosion scales (Schock, Lytle, and Clement 1995; Schock, Wagner, and Oliphant 1996; Schock 1999; Vatankhah et al. 1998). Temperature impacts many parameters that are critical to pipe corrosion including biological activity, physical properties of the solution, thermodynamic and physical properties of corrosion scale, and chemical rates (McNeill and Edwards 2002).

The composition and structure of the surface scales on metals are related to the chemical composition of the finished water. A change in source-water chemistry can destabilize the corrosion scales and expose the metallic surface to further corrosion. Many of the corrosion problems due to change in water chemistry can be related to the change in the structure and composition of surface films and scales (Broo, Berghult, and Hedberg 2001a; Sarin et al. 2001; Tang et al. 2005). Tang et al. (2005) identified the corrosion byproducts occurring on iron surfaces under different source water qualities. The authors report that models based on solubility equilibria can provide a relative measure of iron corrosion byproduct release. However, these models almost always underestimated the iron release due to other complicating factors.

Fransolet et al. (1999) observed that the growth rate of six strains of bacteria in oligotrophic waters was found to be highly dependent on the bicarbonate content of the water. The authors attribute the regrowth of bacteria in nutrient deficient distribution systems to this phenomenon. Since reducing bicarbonate levels is a non-option for most water utilities, the authors stress the need for maintaining proper disinfection. The authors found that phosphate is the limiting nutrient for the bacteria tested. However, Appenzeller et al. (2001) suggests that the phosphate treatment of corroded networks has a positive effect by reducing the release of iron oxide and limiting the proliferation of heterotrophic bacteria in pipes. The risk of bacterial bloom due to phosphate anticorrosion treatment appeared to occur only in some specific waters. Subsequent research by the authors found a reduction of bacterial adhesion on iron oxyhydrides by 75%, indicating that phosphate addition may be beneficial in reducing biofilm formation in old iron pipes (Appenzeller et al. 2002). Butterfield et al. (2002) report that a combination of residual disinfectant and polyphosphate treatment may be required to control biofilms on iron surfaces.

Bade, Manz, and Szewzyk (2000) studied the behavior of sulfate reducing bacteria (SRB) under nutrient deficient conditions and oxygen stress. The authors demonstrated the ability of SRB to survive in nutrient deficient, particle poor habitats like the bulk water in a distribution system. The ability of SRB to survive oxygen stress leads to the possibility of their regrowth in protected areas of the distribution system like biofilms and tubercles. Lytle, Gerke, and Maynard (2005) report that in the anaerobic environment beneath scales of corroding iron pipes, SRB produce sulfide from naturally occurring sulfate in the bulk water. This subsequently can affect the iron geochemistry of the scale as well as the protective qualities of the scale.

Melo and Vieira (1999) studied the physical stability and biological activity of biofilms under turbulent flow (liquid velocity ≥ 1 m/s) and low substrate concentrations. The authors observed that biofilms formed under turbulent flow conditions had increased physical stability and lowered biological activity. Biofilms formed under low velocities had a thick biomass and corresponding higher biological activity.

Frateur et al. (1999) concluded that free chlorine is not electrochemically consumed at the metal surface though it is more easily reducible than oxygen. Therefore, the corrosion of cast iron pipes induces only a chemical decay of chlorine corresponding to its reaction with ferrous ions released by iron dissolution and the total consumption rate is exactly obtained from the corrosion current density. Vikesland and Valentine (2002a; 2002b; 2000) identified similar mechanisms for the consumption of monochloramine by ferrous ions. Decreased microbiological disinfection efficiency can promote the growth of biofilms and subsequent microbiolgically induced corrosion (MIC) problems (Bremer, Webster, and Wells 2001; Critchley, Pasetto, and O'Halloran 2004).

Copper

Copper pipes are commonly used in household plumbing due to their flexibility, ease of installation in plumbing spaces and resistance to thermal stresses. However, copper pipes can corrode and leach unacceptable levels of copper to the drinking water. Copper corrosion is a complex process driven by at least three chemical sub-processes: metal oxidation, fixation of dissolved copper in the corrosion scale and solubility equilibrium. Pitting was often the main failure mechanism identified in copper pipes used for household plumbing (Boulay and Edwards 2001; Harrison, Nicholas, and Evans 2004).

Merkel et al. (2002) described the oxygen consumption (corrosion reaction) by a first order kinetics. The corrosion byproducts identified were cuprite (Cu_2O) and malachite ($CuCO_3.Cu(OH)_2$). The authors report that the copper concentrations observed after extended stagnation periods corresponded with the solubility of malachite phase.

New copper pipes initially release high levels of copper into the water and later the release decreases. This is related to the time required for the formation of stable surface films that subsequently inhibit corrosion. Shim and Kim (2004) observed that a new copper surface was initially covered with cuprous, cupric oxides and cupric hydroxide. Stable cuprous oxide film was formed with increasing immersion time in potable water. Lagos, Cuadrado, and Letelier (2001) observed that langite was the most soluble and youngest film, followed in both solubility and age by cupric hydroxide, azurite, brochantite, malachite, and tenorite. The more stable solids-malachite and tenorite-were 8 and 16 times less soluble, respectively, than the most soluble solid, langite. Scales usually contained more than one compound, a finding that was attributed to temperature changes, variability of water composition, long stagnation periods, and aging.

Harrison, Nicholas, and Evans (2004), observed that pH and buffering capacity were the critical water quality variables needed to sustain malachite scales (thereby reducing increased copper solubility and subsequent pitting). However, the presence of organic matter generally increases copper byproduct release (Boulay and Edwards 2001; Broo, Berghult, and Hedberg 1998; Edwards and Sprague 2001).

Edwards, Hidmi, and Gladwell (2002) report that polyphosphate dosing in high alkalinity waters (300 mg/L) inhibited malachite formation and thereby caused increased copper solubility. Dartmann et al. (2004) observed two effects of phosphate on copper corrosion: Phosphate dosing decreased the copper oxidation and resulted in lower copper concentration in water. Likewise phosphate hindered the precipitation of cupric ions, which prolonged the existence of copper in water and resulted in a higher copper concentration. As a consequence, dosing phosphate can decrease or increase the copper corrosion, depending on the predominant effect.

Hong and Macauley (1998) observed that copper release is determined primarily by residual chlorine, pH and contact time. The authors concluded that the one of the most important parameters is pH since it controls the dissolution of $Cu(II)$ scales. Palit and Pehkonen (2000) observed that the stability of the Cu_2O scale was strongly dependent on the pH, the dissolved inorganic carbon (DIC) and the chloride concentration of the water.

Rushing and Edwards (2004) observed that the orientation and differential heating of a copper pipe can have a significant effect on the amount of copper released into the water due to temperature related changes in solubility of the controlling solid phases of the copper scales. Adeloju and Hughes (1986) observed that the stable product on copper surface up to $50^{o}C$ is $CuCO_3$ and $Cu(II)$ oxides are more stable above $50^{o}C$. The authors attributed this change to increased corrosion and pitting in hot water copper pipes.

Bremer, Webster, and Wells (2001) observed significant biofilm growths to be associated with pit sites on copper pipes and conclude that MIC is also a significant deterioration process for copper pipes. These results were supported by the fact that the most severely corroded pipes also contained the most developed biofilms. Critchley, Pasetto, and O'Halloran (2004) compared the corrosion behaviour of new copper pipes with the introduction of sterile and non-sterile copper corrosion byproducts and observed that corrosion in new copper pipes was higher when non-sterile copper corrosion byproducts were used. The authors concluded that the induction of corrosion and the blue water phenomenon in copper pipes was a result of microbial activity. A previous study by Critchley et al. (2003) reported that cuprosolvency in the presence of biofilms depended on water quality. They reported that the important water quality parameters affecting cuprosolvency in the presence of biofilms are pH, Langlier Index, chloride, alkalinity, total organic carbon (TOC) and soluble phosphate concentrations.

Lehtola et al. (2004a) concluded that new copper pipes inhibit microbial activities based on their observation that the heterotrophic plate count (HPC) in waters was consistently below that of similar PE pipes for the first 200 days. Lehtola et al. (2005b) subsequently reported that copper pipes depleted chlorine residual faster than PE pipes with the result that more chlorine had to be added to water flowing through copper pipes to achieve similar microbial quality to PE pipes.

Copper corrosion of distribution infrastructure can also cause problems downstream at the wastewater treatment plants. Isaac et al. (1997) reported that increased corrosion in water supply systems contributed to the violation of receiving water ambient water quality criteria in one case. Broo, Berghult, and Hedberg (1997) reported a similar case where copper concentration in the wastewater sludge was as high as 600 mg Cu/ kg dry solids.

Lead

Lead in drinking water principally arises from contact between municipally supplied water and lead service pipes (Cardew 2000). Boyd et al. (2001) reported that an estimated 3.3 million lead service lines are located throughout the United States, especially in the older section of major cities. Schock (1999) observed that the permissible lead concentrations in the drinking water of many countries was so low that the rate of lead corrosion was of no practical concern. Even slightly aggressive water in contact with lead or lead containing material will pick up lead after a brief period of stagnation.

Peters et al. (1999) observed that the products formed in the presence of pH adjusted-water were mainly lead carbonate. Products from areas where the water supply had been treated with orthophosphoric acid and pH adjustment for up to eight years also contained a variable proportion of a phosphate species.

Edwards and McNeill (2002) observed that polyphosphate tends to increase release of both particulate and soluble lead to drinking water. In this study, for every milligram per litre of residual hexametaphosphate, soluble lead increased by 1.6 mg/L after 72-hour stagnation in pure lead pipes compared to a control dosed with orthophosphate. The authors recommend that PWSs need to consider these adverse effects whenever polyphosphate is used to prevent scaling or iron precipitation.

Edwards and Dudi (2004) reported a case in Washington, D.C., where a switch of disinfectants from chlorine to chloramines increased released excessive lead concentrations from lead service lines, lead solder and brass plumbing material. Bench-scale testing by the authors found that chlorine reacts with soluble lead to rapidly precipitate a stable lead solid that did not form in the presence of chloramine.

Vasquez et al. (2006) investigated total lead release in the presence of free chlorine and chloramine residuals in drinking waters produced from ground, surface, desalinated, and blended water sources. The authors observed that for both desalinated and blended finished waters, more total lead was released in the presence of chloramines than in the presence of free chlorine.

Deterioration of Polymeric (Plastic) and Bituminous Components

The problem associated with polymeric material is that they leach contaminants (monomers, anti-oxidants and other organics) and are susceptible to the permeation of organic contaminants from the exterior pipe surfaces. These two processes (leaching and permeation) are the major cause of water quality deterioration in the polymeric components of distribution infrastructure. The major water quality factors associated with polymer degradation are residual disinfectant and temperature.

The long-term deterioration mechanisms in polymeric pipes are not documented as well as those of metallic pipes, mainly because these mechanisms are typically slower and because polymeric pipes have been used commercially only in the last 35-40 years. These deterioration mechanisms include chemical and mechanical degradation, oxidation and biodegradation of plasticizers and solvents (Dorn et al. 1996).

Polymer

Age and manufacturing processes can also play a role in polymer degradation. For instance, 'early-era' PVC, manufactured prior to 1977, can leach excessive vinyl chloride monomer into the drinking water (Beardsley and Adams 2003).

Polymeric material contains a variety of organic and inorganic additives to improve the materials durability, the manufacturing and handling, as well as to modify the color (Brocca, Arvin, and Mosbaek 2002). The absorption of water into the polymer matrix may cause swelling or softening. Water can also serve as an extraction agent for the stabilizers and inhibitors, leading to the depletion of these substances by leaching. This causes a concentration gradient that lowers the activation energy for degradation of the polymeric material (Fischer et al. 1993).

Polymeric materials could also impart unacceptable taste and odor (T&O) to the drinking water. Marchesan and Morran (2004) reported that in an evaluation of 314 materials used in drinking water applications, 14 of the 29 products that failed the taste test were polymeric materials. Taste and odor complaints have often led to the identification of gross subsurface contamination.

Based on depth profiling of the deterioration of polymeric material in contact with chlorinated water, many researchers concluded that the chemical species responsible for anti-oxidant loss was highly soluble in the material matrix, while the species responsible for polymer degradation was insoluble since only the immediate surface in contact was degraded (Hassinen et al. 2004; Lundback et al. 2005). This implies that while the structural strength of the polymer is immediately impacted (loss of anti-oxidants leads to increased brittle-failure of polymers), the surface itself may not exhibit extensive signs of deterioration (Albertsson, Andersson, and Karlsson 1987; Fischer et al. 1993). Lundback et al. (2005) reported that even low concentration of residual chlorine (< 0.5 ppm) could decrease the lifetime of polymeric pipes by a factor of 10.

Higher temperatures, as those used in hot water household plumbing exacerbated the deterioration of polymeric components due to rapid loss of anti-oxidants (Gill et al. 1999).

Bonds (2004) evaluated the effects of chloramines degradation on various elastomers used in distribution system fittings. A year-long study by the author concluded that joint gaskets and fittings (with lower exposed surface area) exhibited negligible deterioration compared to sheets of the same material exposed to similar environments. Kilb et al. (2003) observed that biofilms on rubber coated valves seem to act as sources of coliform contamination, whereas no significant biofilm was observed on pipe surfaces.

Polyaromatic Hydrocarbons (PAHs)

Bituminous (coal-tar) linings have been used in the past to limit corrosion of iron and ductile iron pipes and have been implicated in water quality deterioration due to leaching of polyaromatic hydrocarbons (PAHs) (Basu et al. 1987; Maier, Maier, and Lloyd 1999; Maier, Maier, and Lloyd 2000; Merkel et al. 1997; Sorrell, Brass, and Reding 1980).

Bituminous linings contain a complex mixture of many chemical compounds, the exact composition of which depends upon its source. Historically, the use of bituminous coatings has resulted in problems when exposed to soft acidic waters. Among these problems were "red" water discoloration, extensive tuberculation, and loss of hydraulic capacity. Alben, Bruchet, and Shpirt (1989) studied the leaching of organic contaminants from flat steel panels lined with various coatings, including vinyl, chlorinated rubber, epoxy, asphalt, and coal tar. Organic contaminants were found at parts-per-billion levels in water compared to parts-per-thousand levels in the coating. There have been incidents of PAHs leaching from coal tar linings that have ultimately led to T&O complaints. Water quality sampling revealed that anthracene, naphthalene, and other PAHs were present at concentrations in excess of 5000 μg/L (Satchwill 1998).

Five different NSF and AWWA approved epoxy resins were independently investigated by Alben, Bruchet, and Shpirt (1989). Leaching rates were monitored over a period of one month and found to decay exponentially. Concentrations of TOC in leachate from epoxy resin were the highest among two bituminous and three polymeric coatings. The extent of organic contaminants leaching from epoxy resin linings was strongly dependent on the duration of the curing process. Longer curing periods produced more stable linings. The results illustrate that BTEX (benzene, toluene, xylene) and TOC concentrations in stagnant water columns may vary from 0.2-48 mg/L and 34-345 mg/L, respectively, for various epoxy resins assessed over a 30-day period.

Deterioration of Cement-based Components

Cementitious materials in the presence of acidic waters or aggressive waters can degrade in several ways (Leroy et al. 1996). First, the reactions involving hydrolysis and leaching of the components of the hardened cement matrix can increase the porosity and permeability of the component allowing aggressive ions to infiltrate the cement matrix. Second, exchange reactions take place between aggressive fluid and components of the cement matrix resulting in leaching of calcium ions and substitution reactions replacing the calcium in the calcium-silicate-hydroxide (C-S-H) complex resulting in loss of strength and rigidity (softening). Third, the reactions involve the formation of expansive products that can lead to cracking and spalling (Mehta 2003). The extent of leaching increases with water aggressiveness and its residence time in the pipe, which is inversely proportional to pipe diameter. Aggressive waters can dissolve the cement in asbestos-cement (AC) pipes causing the release of asbestos fibers into the water or turbidity from cement pipes and linings (Al-Adeeb and Matti 1984; Guo, Toomuluri, and Eckert 1998; Leroy et al. 1996; Neuberger et al. 1995; Slaats et al. 2004).

Aggressive, soft, and poorly buffered (i.e., low alkalinity) water promotes aluminum leaching from cementitious materials. The USEPA recommends a minimum water pH of 7 in the distribution network, but does not specify minimum alkalinity levels. The same conditions were also found to be conducive to leaching of lead and copper. Douglas, Merrill, and Catlin (1996) studied the deterioration of new cement-mortar linings under various water quality conditions. Field-tests demonstrated that aggressive water is capable of leaching cement compounds from these linings, causing significant increases in pH, alkalinity, and calcium. Asphalt seal coats significantly reduce cement-mortar deterioration.

Furthermore, cementitious materials contain a variety of regulated inorganic chemicals (notably heavy metals), many of which are prone to leaching. Under static conditions, arsenic, barium, cadmium, and chromium leached from the lining reached to within roughly 10-20% of maximum values allowed under the respective drinking water guidelines. Berend and coworkers (1999; 2001) reported a case study of cement-mortar lined ductile iron pipe, which caused elevated aluminum levels of 5 to 690 μg/L during the course of 2 months. Even after two years 100 μg/L aluminum continued to leach from the lining. The study also reports that the leaching imparted several illnesses due to acute aluminum intoxication.

Chemical additives may also affect leaching rates. Berend and Trouwborst (1999) reported polyphosphate (a corrosion inhibitor) reaction on soft cement linings, which accelerated cementitious leaching within average contact time of 2.3 days. The extent of leaching was strongly related to the contact time between the water and the cement-mortar lining.

Maintaining a compatible chemical composition of the finished waters in contact with distribution components can control the rate and magnitude of the deterioration. While this chemical compatibility is easy to achieve for any single material, different components of a distribution infrastructure can have varying responses to change in water quality. Often a change in practice can trigger conflicting results (a positive influence on one component and a negative influence on another). For instance, increasing alkalinity to reduce iron release in distribution systems can cause increased copper release in household plumbing (Broo, Berghult, and Hedberg 1997; Imran et al. 2006). Therefore, managing water quality to control the deterioration of all the components may not be practicable.

RELATIONAL MATRIX (WATER QUALITY VS. MATERIAL IMPACT)

The primary and secondary water quality impacts of the treatment technologies discussed in Chapter 3 are related to their effect on the deterioration of distribution system material. The relational matrix (**C**) (Table 4.4) relates the effect of increase in selected water-quality parameters (WQP) on the deterioration of selected distribution system material (DSM).

Table 4.4
Effect of water quality parameters on different distribution system materials (C)[a]

Parameters	Metals				Cement-Based	Polymeric (Plastic)	
	Iron[b]	Copper	Lead	Galvanized Steel[c]	AC, Concrete[d]	PVC, PE & lining	Bituminous
PH	0.8	0.8		0.8	0.8		
Calcium	0.6	0.5	0.5	0.6	0.6		
Alkalinity	0.8	-0.8	0.8	0.8	0.8		
Sulfate	-0.8	-0.8	-0.1	-0.8	-0.8		
Chloride	-0.8	-0.8	-0.6	-0.8	-0.6		
DO	-0.8[e]	-0.8	-0.8	-0.8		-0.2	
Poly-phosphates	0.8	-0.4[f]	-0.8[g]		-0.5		
Ortho-phosphates	0.8	-0.2	0.8	0.8			
Silicates	0.8[h]	0.8	0.8	0.8	0.8		
TDS	-0.2	-0.2	-0.2	-0.2	-0.2	-0.1	
Ammonia							
Chlorine Residual	-0.4	-0.4	-0.4	-0.4		-0.8	-0.6
Chloramine Residual	-0.2	-0.2	-0.5[i]	-0.2		-0.4	
Natural organic matter	0.5	0.5[j]	0.5[k]		-0.5	-0.5	
Biofilms	-0.8	-0.5	-0.3		-0.6	-0.3	

Effectiveness factor criterion is as follows:

Blank No data

+1.0 Reduced Deterioration — 0 No Effect — -1.0 Increased Deterioration

[a] AwwaRF and DVGW 1996; Schock 1999; Taylor et al. 2005

[b] McNeill and Edwards 2001

[c] Trussell and Wagner 1996

[d] Leroy et al. 1996

[e] Sarin et al. 2004a

[f] Edwards, Hidmi, and Gladwell 2002

[g] Edwards and McNeill 2002

[h] Rushing, McNeil, and Edwards 2003

[i] Lin et al. 1997; Vasquez et al. 2006

[j] Boulay and Edwards 2001; Edwards and Sprague 2001

[k] Edwards and Dudi 2004

The element of the relational matrix (**C**) is a measure of the adverse effect that an increase in WQP_k may cause on DSM_l. Table 4.4 gives an estimate of the magnitude of the impacts that water quality changes will have on common distribution system materials. The impacts identified here are aggregate impacts and should be used for a qualitative evaluation, rather than a quantitative one. For instance, increasing alkalinity can have a dual effect on iron corrosion. It can decrease corrosion by promoting the formation of stable siderite and calcite scales on the iron surface. Conversely, the associated increase in dissolved solids can help in enhancing the electrochemical conductivity necessary for corrosion to occur. However, the overall effect of alkalinity is to reduce the corrosion and release of metallic by-products and is reflected in the effect element (+ 0.8).

CHAPTER 5
EVALUATION OF WATER QUALITY IMPACTS ON DISTRIBUTION INFRASTRUCTURE INTEGRITY

INTRODUCTION

New water quality regulations are a major driver for utilities to consider changing treatment practices. The primary water quality objective is the mitigation of regulated contaminants present in the source water. However, most treatment practices also induce secondary changes in water quality. These often unintentional and sometimes unavoidable secondary water quality changes can have diverse impacts on different materials in the distribution system.

This chapter introduces a framework for identifying relative impacts of changes in treatment processes (to comply with regulations) on some common distribution system materials based on discussion of preceding chapters.

Regulations and Best Available Technologies

The selection of a particular technology for drinking water treatment is governed largely by the PWSs consideration of a number of constraints in addition to the primary requirement of providing safe drinking water. These considerations may include state and national environmental and permitting restrictions on the plant-siting, limitations on withdrawals from a desirable source, restrictions on disposal of process wastes, logistic requirements of monitoring, operation and maintenance, future capacity development and cost of construction (Pontius 2003a).

Under the 1996 SDWA Amendments authorization, the USEPA is required to provide technical and economic guidance to PWSs for compliance with the MCL for each regulated contaminant. To achieve this, the USEPA identifies certain treatment technologies that have demonstrated efficient and economic removal of regulated contaminants. These economically achievable technologies for regulated contaminants are known as 'best available technologies' (BATs).

The BATs are known to work for particular contaminants in some source waters if designed and operated properly. However, due to the diverse source-waters and their treatment requirements, systems have to use their own judgment in the use of the BATs. Though the BATs identified by the USEPA may not always be the best solution to specific water supplies, they provide a useful baseline for the acceptability of available and future treatment technologies (Pontius 2003b).

As a starting point, based on the USEPA's identified BATs, a relational matrix of unit treatment vs. regulations (**A**) was developed, which relates the effectiveness of common unit treatments towards compliance of drinking water regulations. Each element α_{ij} is a measure of the effectiveness of unit treatment '*j*' (*UT*$_j$) towards compliance of drinking water regulation '*i*' (*Reg*$_i$). Matrix **A** was populated with values (Table 3.1), developed based on a comprehensive literature review, where $-1 \le \alpha_{ij} \le +1$. A value of +1 indicates the most effective technology to remove the selected contaminant, while –1 indicates a technology that might adversely affect the removal of the selected contaminant. A value of zero is given to a technology that has no known effect on the removal of a selected contaminant.

$$A = \left(\begin{array}{c|cccc} & UT_1 & UT_2 & \cdots & UT_J \\ \hline Reg_1 & \alpha_{11} & \alpha_{12} & \cdots & \alpha_{1J} \\ Reg_2 & \alpha_{21} & \alpha_{22} & \cdots & \alpha_{2J} \\ \vdots & \vdots & \vdots & \cdots & \vdots \\ Reg_I & \alpha_{I1} & \alpha_{I2} & \cdots & \alpha_{IJ} \end{array} \right) \text{; where } i = 1, 2, \ldots, I \quad j = 1, 2, \ldots, J \tag{5.1}$$

The treatment technologies indicated in Table 3.1 are generic with respect to the contaminant targeted and the mechanisms of contaminant removal. The removal levels attributed to these generic technologies are the maximum removals possible under ideal design and operational conditions.

Best Available Technologies and Secondary Water Quality Effects

The primary objective of any treatment technology is to achieve acceptable removals of the targeted contaminants. However, the use of pre-treatment, treatment and post-treatment practices causes changes in the ionic content of the water and can impact distribution infrastructure.

The matrix **B** relates the effect of unit technology (UT_j) on increase or decrease in selected water quality parameter (WQP_k). Primary water quality changes as well as secondary water quality impacts were evaluated for each specific treatment technology. Matrix **B** was populated with values (Table 3.2), developed based on a comprehensive literature review, where $-1 \le \beta_{jk} \le +1$. A value of +1 indicates a significant increase while −1 indicates a significant decrease in the selected water quality parameter. A value of zero is given to a technology that has no known effect on the selected water quality parameter. For instance, the primary impact of ion exchange is the removal of targeted ions from water and the secondary impact is the removal of non-targeted ions and the addition of counter-ions to the water. Secondary changes induced by the treatment technologies are dose-specific. For instance, depending on the source water quality, different concentrations of treatment chemicals need to be added, and thereby result in different concentrations of secondary ions in the finished water. The water quality parameters selected are those known or suspected in the deterioration of distribution infrastructure.

$$B = \left(\begin{array}{c|cccc} & WQP_1 & WQP_2 & \cdots & WQP_K \\ \hline UT_1 & \beta_{11} & \beta_{12} & \cdots & \beta_{1K} \\ UT_2 & \beta_{21} & \beta_{22} & \cdots & \beta_{2K} \\ \vdots & \vdots & \vdots & \cdots & \vdots \\ UT_J & \beta_{J1} & \beta_{J2} & \cdots & \beta_{JK} \end{array} \right) \text{; where } j = 1, 2, \ldots, J \quad k = 1, 2, \ldots, K \tag{5.2}$$

The magnitude of the secondary water quality changes depends on a number of factors that include; source water contaminants, dosage of treatment chemicals, sequence of treatment practices etc. For instance, though chlorination may be practiced before or after activated carbon filtration, the resulting finished water quality in either case is quite different. Though Table 3.2 provides some inference of treatment impacts on finished water quality, it is not intended as a definitive resource for design purposes. It is used in the context of this report to provide a conceptual evaluation of aggregate water quality impacts of treatment practices.

Water Quality and Distribution System Materials

The water quality impacts on infrastructure surfaces in contact with the finished water are by no means the only (or even the most significant) process that leads to the overall deterioration of the infrastructure components. Other processes including, structural loading, external corrosion, inadequate operation and maintenance and human errors are significant causes of infrastructure failure. However, the water quality induced deterioration may exacerbate the condition of the pipe and make it susceptible to other failure mechanisms.

The relational matrix (**C**) relates the effect of change in selected water quality parameters (WQP_k) on the deterioration of selected distribution system material (DSM). The element γ_{kl} is a measure of the adverse effect that an increase in WQP_k has on DSM_l. Table 4.4 provides values for matrix **C**. These values represent aggregate impacts and should be used for a qualitative evaluation, rather than a quantitative one.

$$C = \left(\begin{array}{c|cccc} & \mathrm{DSM}_1 & \mathrm{DSM}_2 & \cdots & \mathrm{DSM}_L \\ \hline \mathrm{WQP}_1 & \gamma_{11} & \gamma_{12} & \cdots & \gamma_{1L} \\ \mathrm{WQP}_2 & \gamma_{21} & \gamma_{22} & \cdots & \gamma_{2L} \\ \vdots & \vdots & \vdots & \cdots & \vdots \\ \mathrm{WQP}_K & \gamma_{K1} & \gamma_{K2} & \cdots & \gamma_{KL} \end{array} \right) ;\text{ where } \quad k = 1, 2, ..., K \quad l = 1, 2,..., L \qquad (5.3)$$

Table 4.4 gives an estimate of the magnitude of the impacts that water quality changes will have on common distribution system material. The impacts identified here are aggregate impacts and should be used for a qualitative evaluation, rather than a quantitative one. For instance, increasing alkalinity can have a dual effect on iron corrosion. It can decrease corrosion by promoting the formation of stable siderite and calcite scales on the iron surface. Conversely, the associated increase in dissolved solids can help in enhancing the electrochemical conductivity necessary for corrosion to occur. However, the overall effect of alkalinity is to reduce the corrosion and release of metallic byproducts and is reflected in the effect element (+ 0.8).

IDENTIFICATION OF POTENTIAL IMPACTS TO PLANNED CHANGES – DEVELOPMENT OF THE HIERARCHICAL RELATIONAL MODEL (HRM)

This section introduces a method to identify problems related to the integrity of distribution system associated with changing water quality for regulatory compliance. A hierarchical relational model (HRM) is proposed for this purpose. The HRM is schematically illustrated in Figure 5.1. This procedure allows experts in different domains/ fields to provide inputs to the relational matrices depending on their expertise. Also research needs can be identified easily by studying gaps in the effects-matrices.

In Stage 1, the applicable regulations are identified based on the source-water characteristics and contaminants of concern. Using matrix **A** (Table 3.1) as a guide, suitable treatment alternatives are identified to meet the regulations. The composition operation in Stage 1 represents the selection of applicable regulations and treatment alternatives.

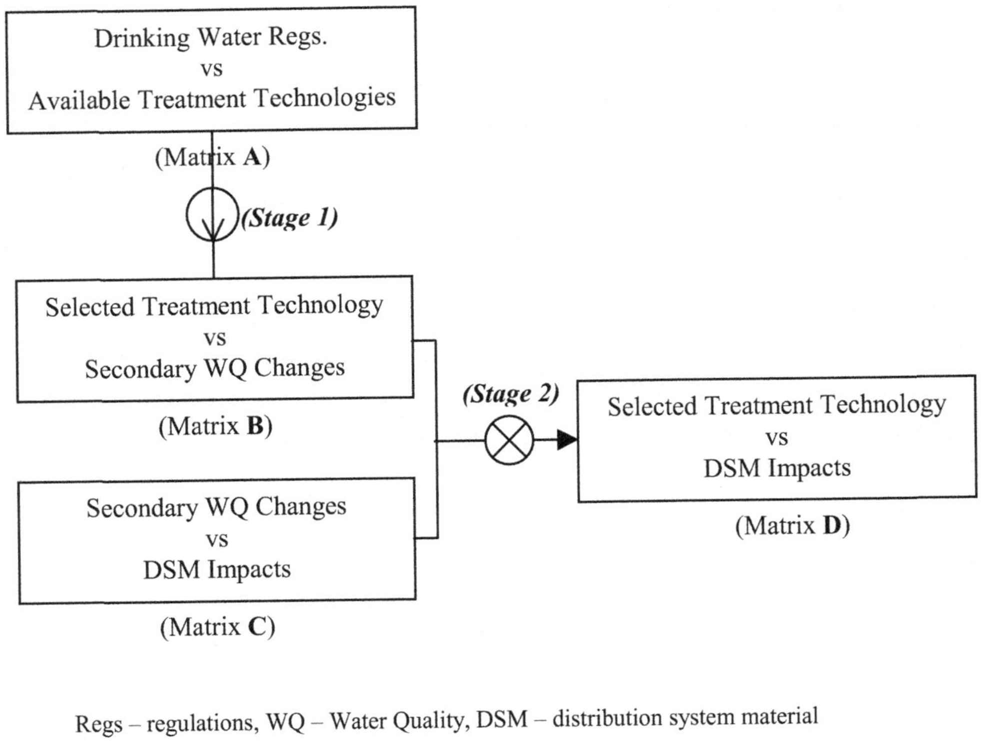

Figure 5.1 Structure of the proposed Hierarchical Relational Model (HRM)

In Stage 2 (Figure 5.1) the composition operation is performed using;

$$D = B \otimes C \tag{5.4}$$

In this case, $\otimes$ represents a composition operation defined by;

$$z_{jl} = \sum_{k=1}^{K} (\beta_{jk} \times \gamma_{kl}), \text{ where } \quad j = 1, 2, ..., J \quad l = 1, 2, ..., L \tag{5.5}$$

The primary and secondary impacts of selected treatment practice are obtained using matrix **B** (Table 3.2). From matrix **C** (Table 4.4), the most common material(s) of construction in the distribution infrastructure are selected. In Stage 2, matrices **B** and **C** are composed to yield matrix **D**, which provides the estimated impact of the selected treatment practice on the distribution materials.

$$D = \left(\begin{array}{c|cccc} & \text{DSM}_1 & \text{DSM}_2 & \cdots & \text{DSM}_L \\ \hline UT_1 & z_{11} & z_{12} & \cdots & z_{1L} \\ UT_2 & z_{21} & z_{22} & \cdots & z_{2L} \\ \vdots & \vdots & \vdots & \cdots & \vdots \\ UT_J & z_{J1} & z_{J2} & \cdots & z_{JL} \end{array} \right); \text{ where } \; j=1, 2, ..., J \;\; l=1, 2,..., L \qquad (5.6)$$

The aggregate impact of the treatment practices, selected in Stage 1, is calculated by simple arithmetic sum of the individual treatment impacts.

$$\hat{D} = \left(\begin{array}{c|cccc} & DSM_1 & DSM_2 & \cdots & DSM_L \\ \hline & \sum_{j=1}^{J} z_{j1} & \sum_{j=1}^{J} z_{j2} & & \sum_{j=1}^{J} z_{jL} \end{array} \right); \text{ where } \; j=1, 2, ..., J \;\; l=1, 2,..., L \qquad (5.7)$$

The algorithm is repeated twice: first for the current (baseline) treatment and second for the changed (alternative) treatment. A difference between the two matrices ($\Delta\hat{D} = \hat{D}_{alternative} - \hat{D}_{baseline}$) will give the relative impact of the water quality change on the distribution system materials. A positive value indicates that water quality "does not adversely affect" the selected distribution material, negative value refers to "adverse impacts" on the selected distribution material, and zero refers to "no change in impact".

Basic Assumptions for HRM

In order to evaluate the aggregate impacts of water quality changes on distribution infrastructure materials, the following simplified assumptions are made:

1. There is a 'linear' relationship between primary and secondary water quality changes and the treatment technique. Linearity in this case implies that a higher level of contaminant in the source-water could require a higher dosage of a treatment chemical and subsequently have higher primary and secondary water quality changes. For instance, this assumption implies that if 2 mg/L of a contaminant requires 4 mg/L of a treatment chemical, then 4 mg/L of the contaminant would require a proportional 8 mg/L of the treatment chemical. In real systems, the dosage is optimized based on pilot or bench-scale studies and may not be linear as assumed.
2. The water quality impacts on distribution system are 'linear' and simple 'additive'. Additive means that the overall water quality impact on any component is the arithmetic summation of all the potential changes due to individual treatment processes. For instance, the net impact on pH of lowering the pH to assist coagulation and subsequently raising pH for stabilization is zero. However, in this case the impact of the secondary water quality change (neutral ions) will persist.
3. The water quality impacts are independent of hydraulic regimes prevalent in the distribution system. This assumption implies that the water quality and hydraulic effects can be separated. However, in actual systems hydraulic effects are also an important consideration in the evaluation of deterioration of the distribution infrastructure, and

4. The water quality impacts are independent of spatial orientation and location of distribution systems. This assumption implies that the water quality impacts are independent of the geometrical shape of the infrastructure component and its location within the distribution system. However, in actual practice, the location of the distribution component (whether it is in a high-flow or a dead-end zone) determines the rate and magnitude of the deterioration.

APPLICATION OF THE HRM

The conceptual framework discussed in the previous section is demonstrated with the help of two hypothetical case studies. Areas of potential concern are obtained by evaluating the relative impact of different alternatives to current (baseline) conditions.

Hypothetical Case Study 1:

Utility X wished to evaluate the impact of changing its secondary disinfectant from chlorine to chloramines to comply with the Disinfectant/ Disinfection Byproduct Rule (D/DBPR) for TTHMs and HAA_5. Though the utility has conducted studies that show that the change will indeed result in compliance with the D/DBPR, there is some concern over the impact on the distribution system material. The utility has identified that iron, copper and lead components comprise more than 90% of its distribution infrastructure.

The first step would be to establish a baseline value for current conditions, to which the proposed changes can be compared. Therefore the rows for TTHM and HAA_5 are activated along with the column for chlorination and chloramination in Table 3.1. It is observed that chloramination has a less adverse DBPs formation impact than chlorination. For the conditions identified, the resulting matrix **A** is:

$$A = \left(\begin{array}{c|cc} & Chlorination & Chloramination \\ \hline TTHM & -1.0 & -0.2 \\ HAA_5 & -1.0 & -0.2 \end{array} \right) \quad (5.8)$$

The secondary water quality changes due to chlorination are given in Table 3.2. Based on the water quality parameters that change during chlorination, the matrix **B** for the current conditions is obtained (5.9).

$$B = \left(\begin{array}{c|ccccccc} & pH & Alkalinity & Chloride & TDS & Ammonia & Residual\ Chlorine & Biofilm\ Formation \\ \hline Chlorination & -0.3 & -0.2 & 0.5 & 0.1 & -1.0 & 1.0 & -0.7 \end{array} \right) \quad (5.9)$$

From Table 4.4, the impact of the water quality on iron, copper and lead is selected (5.10).

$$C = \begin{pmatrix} & Iron & Copper & Lead \\ \hline pH & 0.6 & 0.6 & 0.4 \\ Alkalinity & 0.8 & -0.8 & 0.8 \\ Chloride & -0.8 & -0.8 & -0.8 \\ TDS & -0.2 & -0.2 & -0.2 \\ Ammonia & 0 & 0 & 0 \\ \text{Re}\,sidual\ Chlorine & -0.4 & -0.4 & -0.4 \\ Biofilm\ Formation & -0.8 & -0.5 & -0.8 \end{pmatrix} \quad (5.10)$$

Using Equation (5.5), the resulting effects matrix **D** will be;

$$\hat{D}_{Chlorination} = \begin{pmatrix} & Iron & Copper & Lead \\ \hline Chlorination & -0.60 & -0.49 & -0.48 \end{pmatrix} \quad (5.11)$$

The resulting matrix ($\hat{D}_{Chlorination}$) establishes the baseline for iron, copper and lead deterioration in the distribution system. In order to evaluate the relative impact of chloramination, the relational matrix for chloramination ($\hat{D}_{Chlora\min ation}$) is obtained in a similar manner as was done for chlorination.

$$\hat{D}_{Chlora\min ation} = \begin{pmatrix} & Iron & Copper & Lead \\ \hline Chlora\min ation & -0.22 & -0.37 & -0.82 \end{pmatrix} \quad (5.12)$$

The difference in the two matrices ($\Delta\hat{D} = \hat{D}_{Chlora\min ation} - \hat{D}_{Chlorination}$) gives the relative impact of the change on the distribution system materials (i.e., iron, copper and lead). It is concluded that though the change might benefit iron and copper components of the distribution system, lead components could suffer from increased deterioration.

$$\Delta\hat{D} = \hat{D}_{Chlora\min ation} - \hat{D}_{Chlorination} = \begin{pmatrix} Iron & Copper & Lead \\ \hline 0.38 & 0.12 & -0.34 \end{pmatrix} \quad (5.13)$$

The utility X concludes that further studies are needed to evaluate the impact of the change on lead components of the distribution system.

Hypothetical Case Study 2:

Consider another PWS (Y) that presently practices coagulation with ferric sulfate, followed by sedimentation and filtration. The filtered water is chloraminated prior to supply. However, the utility Y plans to include nanofiltration after the CSF process for further NOM reduction. The utility Y has identified that iron, copper and asbestos-cement are the materials of concern in the distribution infrastructure.

The resulting matrix **B** for the baseline treatment conditions can be obtained from Table 3.2 as;

$$B = \left(\begin{array}{l|ccccccccc} & pH & Alk & SO_4^{2-} & Cl^- & TDS & NH_3 & \text{Residual } NH_2Cl & NOM & \text{Biofilm Form}^n \\ \hline \text{Coagulation} & -0.3 & -0.5 & 0.8 & 0 & 0.1 & 0 & 0 & -0.8 & 0 \\ \text{Filtration} & 0 & 0 & 0 & 0 & 0 & 0 & 0 & -0.6 & 0 \\ \text{Chloramination} & 0 & 0 & 0 & 0.5 & 0.1 & 0.3 & 1.0 & 0 & -0.5 \end{array} \right) \tag{5.14}$$

From Table 4.4, the impact of the water quality on iron, copper and lead is;

$$C = \left(\begin{array}{c|ccc} & Iron & Copper & Asbestos-Cement \\ \hline pH & 0.6 & 0.6 & 0.8 \\ Alk & 0.8 & -0.8 & 0.8 \\ SO_4^{2-} & -0.8 & -0.8 & -0.8 \\ Cl^- & -0.8 & -0.8 & -0.6 \\ TDS & -0.2 & -0.2 & -0.2 \\ NH_3 & 0 & 0 & 0 \\ Res.\,NH_2Cl & -0.2 & -0.2 & -0.1 \\ NOM & -0.5 & -0.5 & -0.5 \\ \text{Biofilm Form}^n & -0.8 & -0.5 & -0.6 \end{array} \right) \tag{5.15}$$

The resulting **D** matrix from the composition shown in Equation (5.5) will be;

$$D = \left(\begin{array}{c|ccc} & Iron & Copper & Asbestos-Cement \\ \hline \text{Coagulation} & -0.8 & -0.04 & -0.9 \\ \text{Filtration} & 0.3 & 0.3 & 0.3 \\ \text{Chloramination} & -0.2 & -0.37 & -0.1 \end{array} \right) \tag{5.16}$$

Using Equation (5.7), the aggregate impacts for the baseline treatment are obtained as;

$$\hat{D}_{CSF-Chloramination} = \left(\begin{array}{ccc} Iron & Copper & Asbestos-Cement \\ \hline -0.76 & -0.11 & -0.72 \end{array} \right) \tag{5.17}$$

Similarly, the aggregate impacts for the addition of membrane process into the treatment trains are calculated.

$$\hat{D}_{CSF-Nanofiltation-Chloramination} = \left(\begin{array}{ccc} Iron & Copper & Asbestos-Cement \\ \hline -0.75 & 1.18 & -1.23 \end{array} \right) \tag{5.18}$$

The relative impact of the change is calculated as;

$$\hat{D}_{CSF-Nanofiltation-Chloramination} - \hat{D}_{CSF-Chloramination} = \left(\begin{array}{ccc} Iron & Copper & Asbestos-Cement \\ \hline 0.01 & 1.29 & -0.50 \end{array} \right) \tag{5.19}$$

This indicates that deterioration of asbestos cement components might be a concern for the utility Y. Based on the results the utility Y evaluates the effect of pH adjustment and the relative impacts are calculated in a similar manner.

$$\hat{D}_{CSF-Nanofiltation-pH\ adjustment-Chloramination} - \hat{D}_{CSF-Chloramination} = \begin{pmatrix} Iron & Copper & Asbestos-Cement \\ \hline 0.77 & 1.73 & 0.45 \end{pmatrix} \quad (5.20)$$

Based on these results the utility decides to conduct further studies to evaluate the effect of proposed changes on asbestos-cement components.

It should be cautioned that the HRM was developed as a concept model and has all the deficiencies inherent in simplifying highly complex relationships (assumptions of linearity and additivity). Though this model was evaluated for common scenarios, the use of this model is dependent on the expertise and experience of a decision-maker. The limitations inherent in the assumptions used to develop this model should temper any use of the model for decision-making.

CHAPTER 6
RESEARCH NEEDS

The identification of research needs for the impacts of water quality on distribution infrastructure is a daunting task due to its complex interrelationships. In order to delineate the impact of water quality, distribution infrastructure is classified based on construction materials of components in contact with drinking water. This project involved following three different avenues for identifying research needs and gaps:

1. Soliciting opinions from a panel of experts through a questionnaire and subsequently discussing the survey results with this panel in a workshop and arriving at a consensus on the survey results and conclusions.
2. Consolidation of expert opinion and literature to identify the research needs at three levels of priorities – *low*, *medium* and *high*.
3. Reviewing existing literature and proposing a simple framework for hierarchical relational model (HRM) to evaluate the impact of changes in treatment processes on the integrity of distribution infrastructure as outlined in Chapter 2 through 5.

EXPERT SURVEY

Expert opinion was solicited in the form of responses to a survey intended to address five issues that were perceived to be important in evaluating water quality impacts on the integrity of distribution infrastructure. Appendix A provides the full text of the survey questionnaire. (The survey was designed as a thinking exercise, which was sent in advance to a panel of experts invited to attend a Research Needs workshop. In addition to the workshop participants, the survey was disseminated to other experts as well and the combined results were presented at the workshop. The participants in the workshop were asked to discuss the survey results and build a consensus on the various issues. Initial responses to the survey were widely variable (primarily as a result of different terminology used by different experts to define infrastructure failure and components). After discussion and several clarifications of terminology that had been ambiguously[a] stated in the survey questions, the panel obtained a convergence of opinions and final results were arrived at by consensus.

State of knowledge

Respondents were asked to identify the current state of knowledge associated with different materials in the distribution infrastructure. Each material had three different categories of associated problems, including: external deterioration, internal deterioration, and related water quality failures. The effect of water quality on the distribution infrastructure is mainly an internal deterioration phenomena, i.e. only surfaces in contact with the drinking water being distributed are subject to deterioration by water quality. However, the actual infrastructure may

---

[a]For instance, experts from Canada had a different outlook on regulatory standards compared to experts from the USA. Similarly, some of the experts in external deterioration of iron pipes had a different view of the term 'lining', than the experts in internal deterioration.

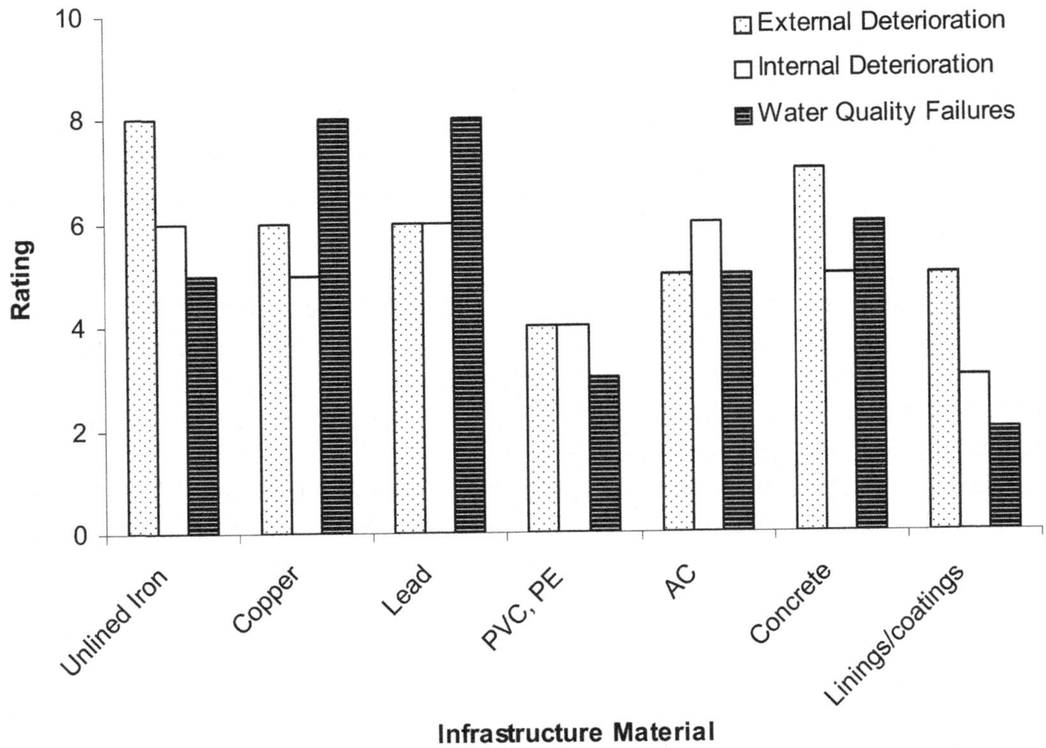

Figure 6.1 Current state of knowledge on different materials based on consensus of participating experts.

fail due to a number of other external factors like improper design, external loading, external corrosion or due to breaks caused by third party interference. The survey participants were also asked to rate from 0 (lowest possible value) to 10 (highest possible value) for the contribution of the internal deterioration processes, if any, on the overall failure of the infrastructure. The participants identified unlined cast iron, copper, lead, PVC/PE and AC components as equally likely to fail as a result of external or internal deterioration. Concrete was perceived as more likely to fail due to external deterioration processes than from internal deterioration (Figure 6.1).

It should be noted that some aspects of the research deficits identified in this report are being undertaken as part of different AwwaRF sponsored research projects. A list of relevant AwwaRF projects with the respective research contractors and the expected dates of completion are given in Table 6.1.

Table 6.1
Ongoing AwwaRF research projects addressing some of the research priorities identified in this study

Project Number	Title	Research Contractor	Expected Completion
2970	Effect of aging water mains on water quality in distribution systems	National Research Council of Canada	2008
2991	Effects of microbes on the leaching of vinyl chloride and organotin from PVC piping used in drinking water delivery	Cornell U.	2007
3018	Contribution of service line and plumbing fixtures to lead and copper rule compliance issues	HDR/EES, Inc.	2008
3100	Secondary impacts of arsenic removal treatment in distribution systems	Malcolm Pirnie, Inc.	2007
3107	Effect of changing disinfectants on distribution system lead and copper release	HDR Engineering, Inc.	2008
3109	Non-uniform internal corrosion in copper piping – Monitoring techniques	Virginia Polytechnic Institute & State U.	2008
3112	Performance and metal release of non-leaded brass meters, components and fittings	HDR Engineering, Inc.	2007
3115	Decision tools to help utilities develop simultaneous compliance strategies	Malcolm Pirnie, Inc.	2008
3118	Assessment of inorganics accumulation in drinking water system scales and sediments	HDR Engineering, Inc.	2009
3172	The role of free chlorine, chloramines and natural organic matter in the release of Lead into drinking water	U. of Iowa	2009
3174	Investigation of the mode of action of Stannous Oxide as an inhibitor of Lead corrosion	U. of Minnesota	2009
4015	Effect of nitrification on corrosion in the distribution systems	Virginia Polytechnic Institute & State U.	2009
4029	Secondary impacts of corrosion control on distribution system and treatment plant equipment	Colorado State U.	2010
4031	Changes in water use patterns	U. of Louisville	2009
4033	Impact of phosphate corrosion inhibitors on cement-based pipes and linings	Camp Dresser and McKee Inc.	2010
4036	Impacts of lining material on water quality	Weston Solutions, Inc.	2010
4064	Influence of water chemistry on the dissolution and transformation rates of Lead corrosion products	Washington State U. in St. Louis	2010

Deterioration processes dominant in various pipe materials

In Chapter 4, three internal deterioration processes were identified as major causes of distribution infrastructure failure (Table 4.3). The participants were asked to identify the most important deterioration process for the different types of material. Each distribution material is susceptible to a different set of environmental stresses that could lead to its failure. The participants identified PVC and other polymeric components as being the least susceptible to physico-chemical or microbially induced corrosion (MIC) (Figure 6.2).

Conversely, unlined iron pipes were identified to be the most susceptible to all three mechanisms of deterioration. However, the results also reflect bias, where the respondents identified problematic material on the basis of water quality deterioration rather than infrastructure deterioration. This is reflected in the respondents' high rating to materials prone to water quality deterioration (unlined iron, copper and lead) (Figure 6.3).

Since the intended focus of this report was on the deterioration of materials (rather than components), respondents were asked to rank the components that were the most likely to fail as a result of the impact of water quality. The respondents indicated that the most vulnerable component was the household plumbing and fixtures as well as small diameter mains and connections, followed by pump and pump appurtenances (Figure 6.4).

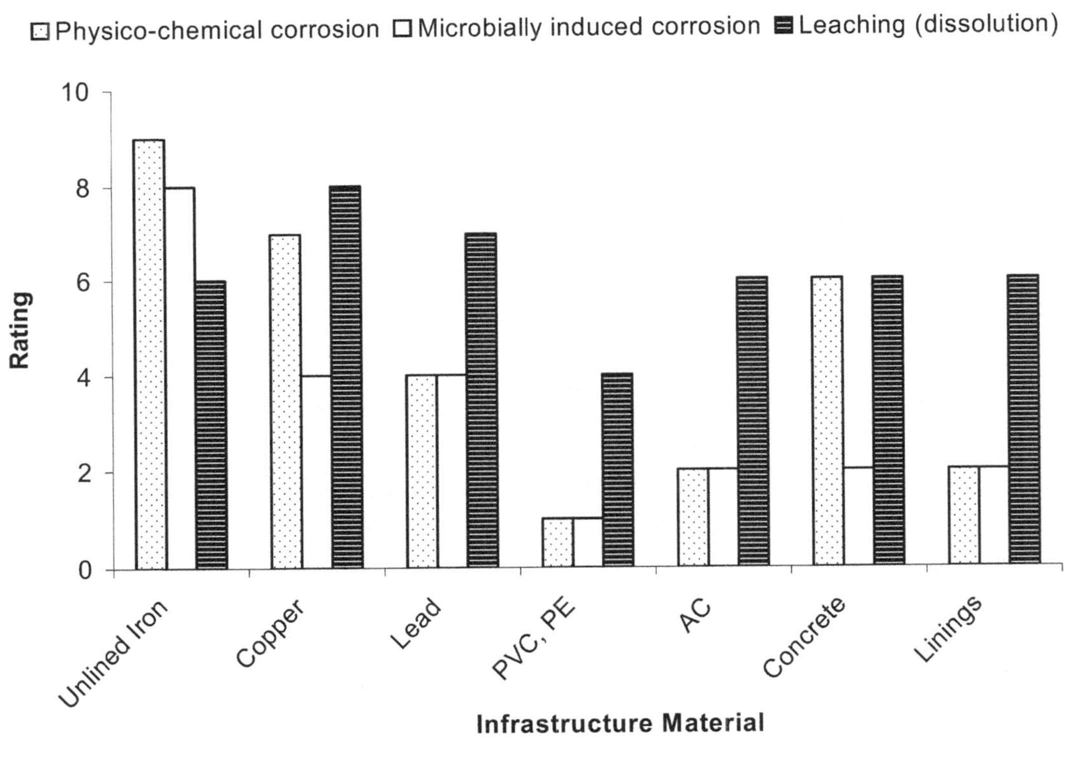

Figure 6.2 Significant deterioration processes for different materials according to survey results.

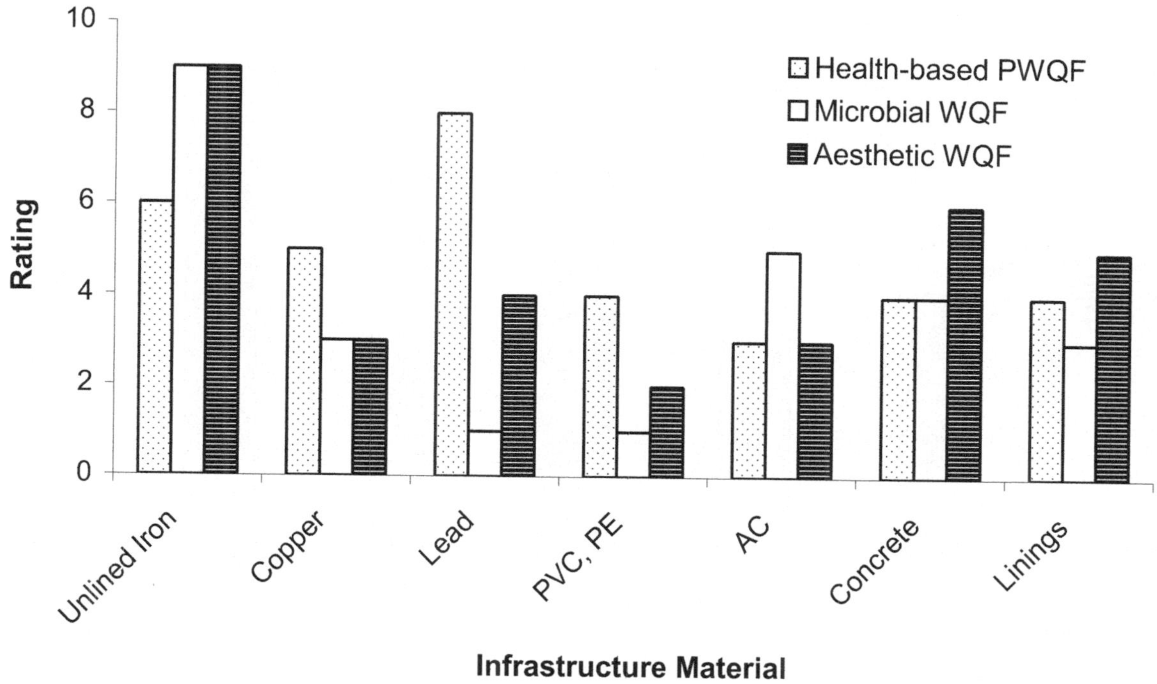

Figure 6.3 Water quality failures associated with different materials

Identification of distribution pipes condition

One of the main challenges facing the drinking water purveyors is identifying the state of internal deterioration of their existing distribution infrastructure. The direct inspection of the condition of the inner surfaces of distribution mains is a rather costly prospect. Practitioners therefore often have to resort to indirect indicators that might provide evidence on the condition of these pipes. The participating experts were asked of their opinion on the different methods of determining the internal condition of a pipe and the state of knowledge associated with these methods. Electrochemical techniques such as the use of corrator probes for corrosion monitoring, received the lowest rating in both applicability and state of knowledge. High ratings were given to water quality monitoring, water leakage and inspection of exhumed pipes. It should be noted that the experts tended to rate demonstrated technologies higher than newer ones. This could be used to indicate that electrochemical techniques remain an area for further research and development in drinking water infrastructure (Figure 6.5).

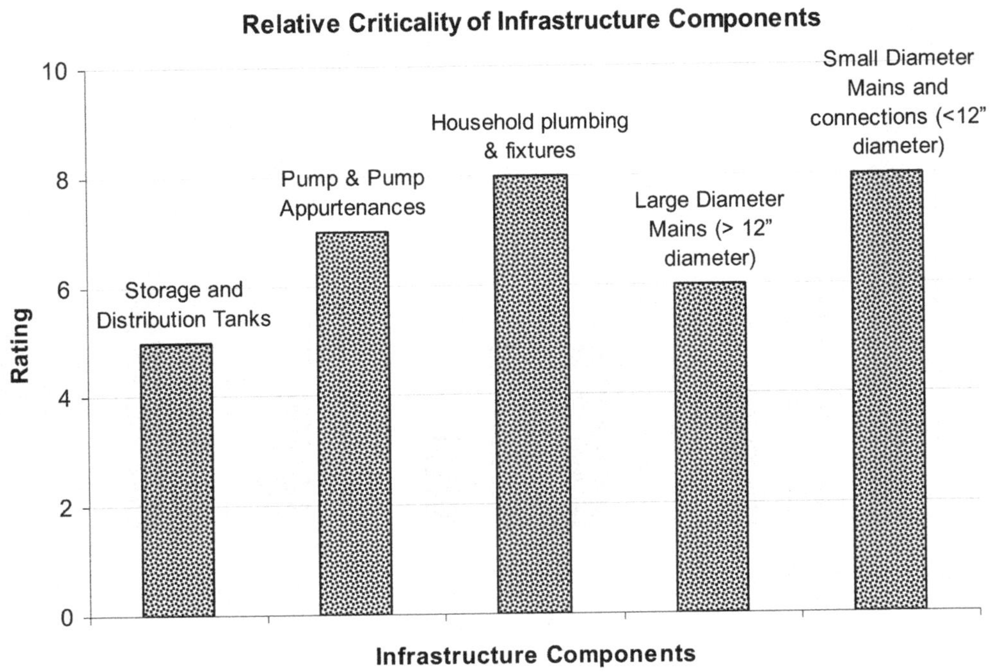

Figure 6.4 Relative importance of components based on water quality impacts

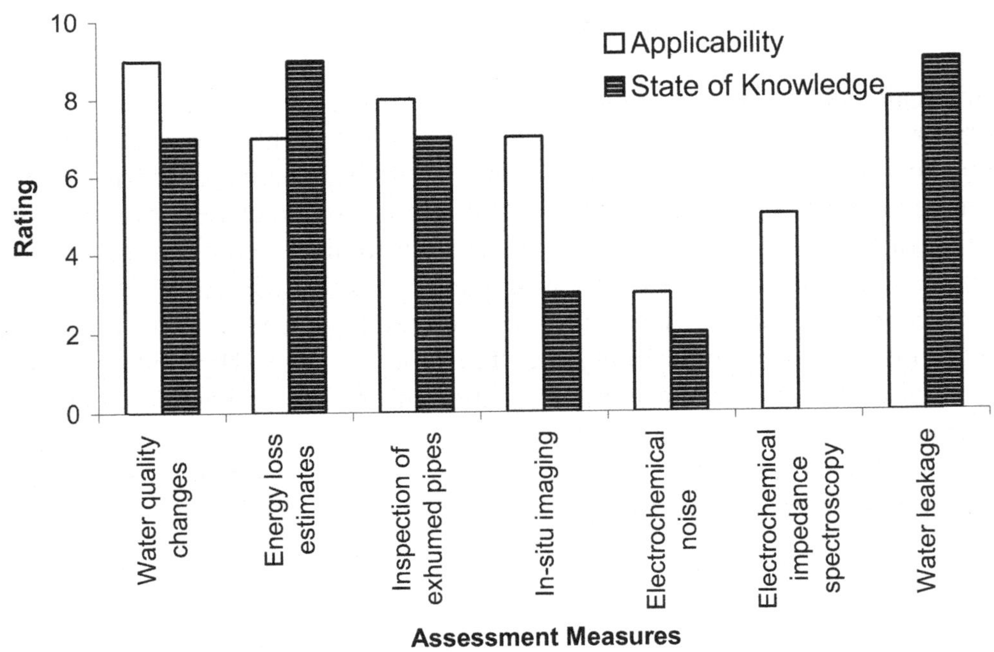

Figure 6.5 Rating of assessment measures for integrity of distribution infrastructure

RESEARCH PRIORITIES IDENTIFIED IN THIS STUDY

Participating experts were given a list of research foci for each pipe material and asked to identify the most important ones. The participants were free to add their own ideas to the list, and an item was added to the list based on a suggestion from a participant. The research issues (and needs) were classified by consensus as *high*, *medium* or *low* priority. The resulting prioritized list of issues is suggested as a basis upon which new research projects could be defined, based on both water quality as well as mitigation of deterioration of distribution infrastructure.

Research Priorities for Unlined Iron Components

A major portion of a typical water distribution network in North America comprises of unlined cast iron pipes that are at least a few decades old. Table 6.2 identifies research priorities for unlined cast iron components. This category includes old galvanized pipes and iron pipes with insufficient lining. Areas of current concern in unlined iron pipes include the accumulation and release of heavy metals like arsenic, lead, cadmium etc., from the internal surfaces. There is also a need for an assessment tool for measuring the internal condition of old pipes. This would help in planning rehabilitation activities.

Research Priorities for Lead and Copper Components

The promulgation of the LCR has provided an impetus towards the mitigation of lead and copper release. However, the conflicting impacts of water quality and treatment changes on distribution infrastructure are not fully understood. Lead release due to chloramination is an example of how changes intended to comply with one regulation (Stage 2 D/DBPR) can conflict with the compliance of another (LCR) (Vasquez et al. 2006; Edwards and Dudi 2004). Table 6.3 provides the identified research needs for lead and copper in the distribution system.

It should be noted that health is the primary concern with lead pipes and connections due to ingestion of lead corrosion byproducts. Therefore, the most feasible alternative for lead abatement is replacement of lead-based components in the distribution infrastructure by components made with more acceptable materials. Protection of lead pipes from deterioration is advised only in cases where the deterioration causes lead release to the drinking water.

For lead and copper pipes a major area of concern is the galvanic corrosion of these materials at the interface where they are in contact with a newer pipe material. For copper pipes, there is an additional need to evaluate whether passivated old copper pipes and newly installed copper pipes exhibit any significant galvanic corrosion.

Research Priorities for Polymeric Components

The polymeric pipes and components (PVC, CPVC, HDPE, PE, elastomers) group is the most diverse and understudied in the distribution infrastructure. The excellent corrosion and microbiological resistance offered by these pipes makes them the fastest growing replacement choice for water utilities. However, given that the average life of a pipe in the distribution system is measured in decades, more research is needed on the long-term performance of polymeric pipes. Table 6.4 identifies some of the research needs identified for this pipe group.

Table 6.2
Research priorities identified for unlined iron in the distribution infrastructure

Priority	Impacts on Distribution Infrastructure	Impacts on Water Quality
HIGH	*† Develop and test methods of evaluating the impacts of different disinfectants on deterioration of old pipes. § Control of corrosive action of common disinfectants. *† Identify interactions between corrosion rate, microbiological activity and disinfectant type and residual.	‡ Develop and test methods for comparison of disinfection efficacy for different pathogen groups using different disinfectants. *† Develop models for residual disinfectant depletion due to interaction with corrosion byproducts and pipe walls. ‡ Model and evaluate accumulation, growth and release of physical, chemical and microbiological contaminants.
MEDIUM	‡ Document and model the genesis and evolution of corrosion scale and release controlling solid phases under different water qualities. * Develop guidance for mitigation of internal deterioration in old pipes. § Document and compare performance of newly lined iron pipes on remaining life of pipe.	*† Develop practical solutions for water quality deterioration.
LOW	‡ Develop a forensic model to identify internal deterioration problems based on the type and concentration of the particulate released. ‡ Evaluate accumulation and release of heavy metals from corrosion scales.	§ Evaluate control measures to increase the formation of passivating solid phases.

* Some research needed; ‡ Considerable research done; § No research done yet; † continuing research in active AwwaRF projects

Table 6.3
Research priorities for Lead and Copper in the distribution infrastructure

Priority	Impacts on Distribution Infrastructure	Impacts on Water Quality
HIGH	*† Develop and test methods of evaluating the impacts of different disinfectants on deterioration of copper pipes. § Control of corrosive action of common disinfectants. ‡ Identify interactions between corrosion rate, microbiological activity and disinfectant type and residual.	‡ Develop models for residual disinfectant depletion due to interaction with pipe walls. ‡ Evaluate impacts of chloramines on lead and copper release and LCR compliance. ‡ Evaluate impact of new copper plumbing installations on LCR compliance. ‡ Evaluate impact of partial replacement of lead pipes on LCR compliance.
MEDIUM	‡ Document and model the genesis and evolution of the corrosion scales and release controlling solid phases for lead and copper under different water qualities. ‡ Document and compare corrosion of newly installed and old copper pipes.	* Develop practical solutions for water quality deterioration in both lead and copper pipes.
LOW	§ Develop a forensic model to identify internal deterioration problems based on the type and concentration of the particulate released from both lead and copper pipes.	§ Evaluate control measures to increase the formation of passivating solid phases in both lead and copper pipes.

* Some research needed; ‡ Considerable research done; § No research done yet; † continuing research in active AwwaRF projects

PVC, PE and other plastic components represent a newer and continuously evolving material used in the distribution systems. These materials have a limited history and therefore need further research. Specifically, as shown in Table 6.4, the areas of concern are the leaching of different organic components into the drinking water and the degradation of these materials through oxidants (mainly disinfectants) in the drinking water.

Concerns about polymeric pipes mostly arise in hot water piping systems, where the combination of high temperatures, residual disinfectants and other oxidants could cause a deterioration of the pipes. On the other hand, many polymers are also capable of withstanding high temperatures and disinfectant concentrations. Therefore when selecting polymeric pipes a careful evaluation of its compatibility with the operating and environmental conditions should be conducted.

In the survey, the 'state of knowledge' concerning the deterioration of polymeric components was ranked the lowest by the participating experts (Figure 6.1). This implies more research is needed to study the long-term behavior of polymeric pipes.

Table 6.4
Research priorities for PVC, PE and other plastics in the distribution infrastructure

Priority	Impacts on Distribution Infrastructure	Impacts on Water Quality
HIGH	‡ Identify and control permeation of organic contaminants through pipe walls of plastic pipes. †‡ Impact of changing disinfectants on elastomers and other plastic components of the distribution infrastructure.	§ Evaluate impacts of chloramines on leaching from pipe matrix. ‡ Characterization of contaminants leaching from plastic pipes and their impact on health and aesthetic water quality.
MEDIUM	‡ Examine and quantify the leaching of monomers and other organic components from the pipe matrix.	‡ Develop practical solutions for water quality deterioration in plastic pipes.
LOW		‡ Evaluate impacts of new plastic installations on health and aesthetic water quality deterioration.

* Some research needed; ‡ Considerable research done; § No research done yet; † continuing research in active AwwaRF projects

Research Priorities for Cement-based Components

Two major categories of cementitious material exist in the distribution infrastructure, Asbestos cement and concrete. Since the cement matrix in both materials is pozzolanic in nature, a similar reaction to water quality can be expected in both. However, the mode of deterioration in each could be completely different. In concrete, structural deterioration is due mainly to the corrosion of pre-stressing steel wires and subsequent loss of tensile strength. Other deterioration due to cracking and spalling could occur due to reaction of the salts within the cement matrix. Research priorities for cement-based material are identified in Table 6.5.

Research Priorities for Internal Linings

Table 6.6 gives some of the research priorities identified for lining material in the distribution systems. These liners are typically used either as a component of new pipes or for rehabilitation of old pipes. In both cases, the long-term performance of the liners needs to be evaluated.

Linings are often the rehabilitation measure of choice for unlined iron (and other) pipes. The liner material in effect becomes the surface in contact with the drinking water. The water quality impact in this case could occur in two stages, slow degradation of the liner material and exposure of the base underlying metal followed by intense localized corrosion of the base metal. Therefore it is necessary to study the long-term performance of different lining materials.

Table 6.5
Research priorities for AC, Concrete and other cementatious materials in the distribution infrastructure

Priority	Impacts on Distribution Infrastructure	Impacts on Water Quality
HIGH		‡ Model and evaluate accumulation and release of heavy metals from pipe surfaces.
MEDIUM	‡ Evaluate softening of pipe materials under different water quality scenarios. § Evaluate impact of disinfectants on integrity of pipe.	‡ Evaluate impact of new installations on health and aesthetic water quality deterioration.
LOW	*† Corrosion of reinforcing steel or prestressing wires and impact on integrity of concrete structures and pipes.	

* Some research needed; ‡ Considerable research done; § No research done yet; † continuing research in active AwwaRF projects

Table 6.6
Research Priorities for liners (polyaromatic hydrocarbons, epoxy, etc.) in the distribution infrastructure

Priority	Impacts on Distribution Infrastructure	Impacts on Water Quality
HIGH	‡ Evaluate the long-term performance of lined pipes and liners.	
MEDIUM	‡ Leaching of PAHs, monomers and other matrix components from liners and their impact on the durability and effectiveness of linings.	‡ Evaluate impact of new installations on health and aesthetic water quality deterioration. ‡ Evaluate the long-term leaching of contaminants from liners.
LOW	‡ Impact of disinfectants on performance of liners.	

* Some research needed; ‡ Considerable research done; § No research done yet; † continuing research in active AwwaRF projects

Other Research Priorities for Distribution Infrastructure

Other research priorities identified in Table 6.7 address water sources, water treatment and operation, maintenance and renewal activities, and are intended to minimize disruption to the existing distribution infrastructure. There is also an urgent need for addressing scale-up issues in laboratory and bench-scale experiments as they apply to the distribution system.

Different utilities have different research needs based on the applicability of specific regulations, and use of different treatment processes and materials of constructions. The authors have proposed an innovative but simple framework (model) by which utilities can evaluate the impact of changes in treatment processes to the existing infrastructure. Different materials have different, often opposing water quality requirements to mitigate deterioration. Conflicting water quality impacts on different distribution materials can also be identified using this model. The model proposed in this project is intended as a proof of concept and can be further refined to incorporate more complex real systems. Further development of similar assessment models to aid decision-making in evaluating treatment changes as well as evaluating material for the replacement of existing infrastructure is needed.

Table 6.7
Other research priorities for distribution infrastructures

Priority	Impacts on Distribution Infrastructure	Impacts on Water Quality
HIGH	‡ Develop and evaluate methods to identify water quality impacts on different materials simultaneously. ‡ Develop a standard assessment procedure to identify the extent of internal deterioration in pipes and other inaccessible infrastructure. ‡ Develop methods to evaluate scale-up issues arising from bench and laboratory studies.	‡ Develop methods to evaluate and optimize water sources and treatment processes to protect the existing distribution infrastructure.
MEDIUM	‡ Determining optimum monitoring strategies to evaluate internal deterioration in distribution systems. ‡ Mechanisms of inhibition by corrosion inhibitors.	‡ Develop methods to evaluate scale-up issues arising from bench and laboratory studies.
LOW	‡ Impact of disinfectants on performance of liners.	

* Some research needed; ‡ Considerable research done; § No research done yet; † continuing research in active AwwaRF projects

CHAPTER 7
SUMMARY OF RECOMMENDATIONS

The main objective of this research was to identify research needs for evaluating the impact of water quality on the integrity of distribution infrastructure. This was accomplished by an extensive review of current literature, development of a framework for hierarchical relational model (HRM), and expert opinion through a survey and workshop.

Each water utility has it own set of challenges based on the source water used, demand dynamics, applicable regulations, condition of distribution infrastructure, operation, maintenance and renewal strategies, economic and management issues. The HRM proposed in this research allows each utility to predict the relative impact of changing treatment process and practices for its unique conditions. A further refinement and evaluation of the HRM is recommended.

Experts from various disciplines in drinking water were polled for their opinion through a survey. The results of this survey were discussed during a research needs workshop and a consensus was built around the most and the least pressing research needs for different materials of construction of the distribution infrastructure. The detailed description of the research areas and their priorities identified in this project is given in Table 6.1 to Table 6.6.

The following gives a general idea of the knowledge gaps and arising research needs identified in this research project:

1. The evaluation of how changes in water sources (and qualities) can affect different components of the distribution network; for instance, some surface waters can corrode unlined iron pipes faster than groundwater;
2. The reconciliation of conflicting impacts of water quality (chemistry) on different pipe materials; for instance, water composition compatible with iron pipes may not be compatible with copper pipes. Therefore, a change in water composition, by the addition of chemicals to protect iron pipes, may cause failure of copper pipes;
3. The evaluation of unintended effects of change to water treatment processes; for instance, increasing the amount of chlorine to kill pathogenic microorganisms can cause deterioration in some pipe materials;
4. The evaluation of the effect of changes in water composition on the host pipes in terms of both the duration and frequency of these changes. For instance, the effect on distribution pipes when the utility regularly switches between surface water and groundwater sources to meet seasonal demand; and
5. The development of a decision-support model (e.g. HRM) to help water utilities evaluate the effect of changing water treatment practices on the deterioration of different pipe materials.

APPENDIX A
SURVEY USED FOR THE RESEARCH NEEDS WORKSHOP

National Research Council Canada
Institute for Research in Construction
Ottawa, Canada
K1A 0R6

Conseil national de recherches Canada
Institut de recherche en construction

NRC · CNRC

March 8, 2006

Dear Colleague,

The NRCC-AwwaRF collaborative project titled *"Examining the impact of water quality on the integrity of distribution infrastructure"*, aims at establishing current trends and future research needs. The main focus of this study is to identify how the water distribution network might be impacted by water quality changes that occur when water utilities attempt to comply with new regulations.

Thank you for volunteering your time and expertise for the NRCC-AwwaRF research-needs survey. The results of this survey will be disseminated in the form of an AwwaRF report, followed possibly by journal and conferences publications.

The names and affiliations of all participants in the survey will be acknowledged in the final AwwaRF report. You can e-mail (preferred), fax or mail the completed survey to us.

We would appreciate your input for this important and timely research by *25th March 2006*. We would appreciate it if you could acknowledge your participation as soon as possible. Please feel free to contact us in case you need additional information.

Thank you,

Dr. Rehan Sadiq
Research Officer
National Research Council, Canada
1200 Montreal Rd., Bldg. M-20
Ottawa, ON, K1A 0R6, Canada
Tel: 1-613-993-6282
Fax: 1-613-993-1866
E-mail: rehan.sadiq@nrc-cnrc.gc.ca

Dr. Syed A. Imran
Research Visiting Fellow
National Research Council, Canada
1200 Montreal Rd., Bldg. M-20
Ottawa, ON, K1A 0R6, Canada
Tel: 1-613-993-0122
Fax: 1-613-993-1866
E-mail: syed.imran@nrc-cnrc.gc.ca

BRIEF OUTLINE OF THE NRCC-AwwaRF COLLABORATIVE PROJECT

Utilities, in the United States, increasingly report that their treatment system upgrades are driven by the need to comply with new and forthcoming United States Environmental Protection Agency (USEPA) regulations (AWWA 2000). These treatment changes lead to corresponding changes in the water quality and ultimately impact the water distribution infrastructure. The need arises to evaluate common compliance practices for their impact on water quality and subsequently on the distribution infrastructure. The main objective of this project is to identify research-needs for evaluating water quality impacts on distribution infrastructure integrity. This is achieved by identifying the missing links in the complex interrelation between regulations, required treatment technologies to comply with these regulations, secondary impacts on water quality and associated distribution material impacts.

RESEARCH OBJECTIVES

The present study aims at evaluating the USEPA drinking water regulations as a driver for change in the condition of distribution infrastructure. This task will be accomplished by;

- review of current and proposed USEPA drinking water quality regulations,
- review of common treatment technologies for compliance and the associated water quality,
- review of impact of water quality on common distribution materials, and
- identify the research needs and knowledge gaps to establish links between the changing water quality regulations and compliance strategies on distribution integrity.

STATE OF PRACTICE

Utilities today are under constant pressure to provide safe drinking water to consumers in acceptable quantity and quality. Increasing regulatory restrictions on the source-water withdrawals, treatment techniques and maximum contaminant levels (MCLs) require utilities to constantly change, upgrade or eliminate their historic water sources and treatment practices. These strategies to comply with the state and national regulations on the drinking water quality and quantity ultimately lead to changes in the quality (chemistry) of the water that comes in contact with the distribution networks. Though there is a general awareness that these changes might impact the distribution infrastructure adversely, utilities have historically taken a myopic outlook on protection of distribution infrastructure through water quality.

USEPA DRINKING WATER REGULATIONS

Maintaining regulatory standards of water quality is one of the major drivers of change in utility practice and subsequently in water chemistry (quality). Therefore, it is important to examine the current and proposed changes in drinking water regulations that might potentially change the water quality. Recent rulemakings by the USEPA have increased in complexity and scope. Promulgations of the Disinfectant/Disinfection Byproduct Rules (D/DBPRs), Surface Water Treatment Rules (SWTRs), Groundwater Rule (GWR), Total Coliform Rule (TCR), and Lead and Copper Rule (LCR) indicate that the USEPA increasingly favors a managed approach to drinking water safety rather than an MCL at the distribution entry approach. However, in spite of evidence that distribution system deficiencies are among the major causes of waterborne outbreaks, a comprehensive regulatory approach to distribution integrity has not been attempted.

Any managed attempts at controlling contaminants (including pathogens) are as good as the state of the distribution system. Even complete compliance with all the regulations does not guarantee safe drinking water unless the distribution infrastructure is also in a good condition.

WATER TREATMENT TECHNOLOGIES

In the selection of water treatment technologies utilities have to balance the needs of present and future expansion, local and national environmental restrictions on siting, source-water usage and disposal of waste streams. These objectives are in addition to the primary objective of producing safe drinking water. Utilities need to pay particular attention to the placement or sequence of these treatment changes. Secondary water quality changes should also be evaluated to determine potential problems related to adverse impacts on distribution infrastructure.

DISTRIBUTION INFRASTRUCTURE IMPACTS

'Integrity' of distribution infrastructure can be defined as its ability to transport water in acceptable quantity and quality without causing any 'structural' or 'functional' failure of its components.

Distribution infrastructure is a complex amalgamation of pipes and appurtenances constructed of different materials, of different age and different manufacturing processes. Changes in water quality can potentially impact the distribution infrastructure by causing internal corrosion-related problems. Internal corrosion and related problems in the distribution infrastructure are unique in the sense that the extent of deterioration is governed by the chemistry of the water being transported, while at the same time the chemistry of the water is impacted by the corrosion of the internal surfaces of the pipes. While the water/internal surface interactions are inextricably interrelated, a plethora of literature exists that examines the water quality deterioration due to the interaction with internal pipe surfaces. Unfortunately, the converse phenomena of deterioration of the internal pipe surfaces due to the water quality has only been indirectly studied.

PERSONAL INFORMATION[a]

Name	
Designation	
Institution/Company	
Contact Address	
Phone No:	
Fax No.	
E-mail Address	

Include Name and Affiliation in final AwwaRF report? Yes _____ No _______

Comments:

Please see footnotes for definitions and help on filling the survey.

You can add any relevant information in the space provided.

[a] The contact information will only be used in case further clarification to the survey input is needed from you.

SURVEY

The deterioration of distribution infrastructure[a] is mainly due to by 'external' and 'internal' processes (mechanisms). Internal deterioration is due to the interaction of drinking water with surfaces in contact. External deterioration is due to external corrosion (physico-chemical and bio-chemical), fatigue, etc. It is understood that both internal and external deterioration may play a synergistic role in the overall deterioration (which could lead to compromised integrity of distribution infrastructure). However, the exact contribution of each towards overall deterioration is unknown. In question 1, we attempt to elicit expert opinion on the relative impact of these two deterioration types on the overall 'deterioration'.

1. **Please rate on a scale of 0 to 10 the contribution of internal and external processes towards overall deterioration of the distribution infrastructure?[b]**

(Scale 0 – 10, 0 → no contribution, 10 → significant contribution, N: not sure)

Row / Column	External Deterioration	Internal Deterioration	Others (specify)
Unlined Iron			
Copper			
Lead			
PVC & PE			
Asbestos Cement			
Concrete			
Linings (specify type) ________			
Other (specify) ________			
Other (specify) ________			

Comments:

[a] In this survey, the terms 'distribution infrastructure', 'water distribution network', 'water distribution system', 'distribution network' and 'distribution system' are used interchangeably to mean 'water distribution infrastructure', including pipes and appurtenances (valves, fittings).

[b] For the deterioration (Row 1) and materials (Column 1), fill in the appropriate number indicating your opinion on the contribution of the deterioration process on the overall deterioration

'State of knowledge' is an aggregate measure of technical and scientific awareness in terms of established practices and current research foci.

2. Please rate the current 'state of knowledge'?[a]

(Scale 0 – 10, 0 → no knowledge, 10 → significant knowledge, N: not sure)

Row / Column	External Deterioration	Internal Deterioration	*Water quality failures	Others (specify)
Unlined Iron				
Copper				
Lead				
PVC & PE				
Asbestos Cement				
Concrete				
Linings				
Other (specify) ____________				
Other (specify) ____________				

* In this survey, the term 'water quality failure' is referred to violation of regulations (or guidelines or self imposed limits) of one or more water quality indicators.

Comments:

[a] For the deterioration process (Row 1) and materials (Column 1), fill in the appropriate number indicating your opinion on the current state of knowledge/research.

In our opinion there are three dominant internal deterioration mechanisms; (physico-chemical based) internal corrosion, microbiologically induced corrosion (MIC) and leaching (dissolution) of material. Though any material may have a specific predominant mechanism of deterioration, all three processes could play some role towards its overall deterioration.

3. **Please rate the impact of the following *internal* deterioration mechanisms on overall *internal* deterioration?**[a]

(Scale 0-10, 0 → no role, 10 → significant role, N: not sure)

Row / Column	Physico-chemical based corrosion	Microbiologically induced corrosion	Leaching (dissolution)	Others (specify)
Unlined Iron				
Copper				
Lead				
PVC & PE				
Asbestos Cement				
Concrete				
Linings				
Other (specify) ______				
Other (specify) ______				

Comments:

[a] For the internal deterioration mechanisms (Row 1) and materials (Column 1), fill in the appropriate number indicating your opinion on the role of the deterioration mechanism towards overall internal deterioration.

The 'water quality failures' in distribution systems can be categorized into, physico-chemical (e.g. disinfection by-products, contaminant release etc.), microbiological (e.g. pathogens, biofilms etc.) and aesthetic (e.g. taste & odor, color etc.) failures.

4. **Please rate the extent of water quality failures associated with *internal* deterioration of pipes?**[a]

(Scale 0-10, 0 → not significant, 10 → highly significant, N: not sure)

Row / Column	Physico-chemical water quality failure	Microbiological Water quality failure	Aesthetic water quality failure	Others (specify)
Unlined Iron				
Copper				
Lead				
PVC & PE				
Asbestos Cement				
Concrete				
Linings				
Other (specify) ____________				
Other (specify) ____________				

Comments:

[a] The extent of water quality failure (Row 1) associated with the material (Column 1)

The primary objective of any treatment technology is to achieve acceptable removals of the targeted contaminants. However, the use of pre-treatment, treatment and post-treatment chemicals changes the chemical composition of the water and subsequently impact different deterioration processes.

5. Please rate the impact of the following water quality changes on *internal* deterioration processes?[a]

(Scale 0-10, 0 → no role, 10 → significant role, N: not sure)

Row / Column	Inorganic Composition	Organic Composition	Disinfectant (specify type)	Dissolved gases	Other (specify)
Physico-chemical corrosion					
Microbiologically induced corrosion					
Leaching (dissolution)					
Other (specify) ____________					
Other (specify) ____________					

Comments:

[a] The effect of change in water quality (Row 1) on the internal deterioration processes (Column 1)

A major portion of the distribution infrastructure is inaccessible. This makes assessment of the state of internal deterioration difficult.

6. Rate the following indicators (or methods) for applicability (or suitability) and 'state of knowledge' to assess the extent of *internal* deterioration?

(Scale 0-10, 0 → negligible, 10 → high, N: not sure)

Row / Column	Applicability	State of Knowledge
Changes in water quality (e.g., color) within the distribution network		
Energy loss estimates (due to change in roughness)		
Inspection of exhumed pipes		
In-situ imaging technologies		
Electrochemical noise analysis		
Electrochemical Impedance Spectroscopy		
Water loss estimates		
Other (Specify) ______________		

Comments:

7. **Please rate 'criticality' of each of these distribution components with respect to impact of change in water quality.**

(Scale 0-10, 0 → not critical, 10 → Critical, N: not sure))

Components	Rating
Storage and Distribution Tanks	
Pump & Pump Appurtenances	
Household plumbing & fixtures	
Large Diameter Mains (> 12" diameter)	
Small Diameter Mains and connections (< 12" diameter)	
Other (specify)	
Other (specify)	

Comments:

New US water quality regulations are a major driver for utilities to consider changing treatment practices. The primary water quality objective is the mitigation of regulated contaminants present in the source water. However, most treatment practices also induce a secondary changes in water quality. These often unintentional and sometimes unavoidable secondary water quality changes can have diverse impacts on different materials in the distribution system.

8. Rate the potential water quality impacts of compliance with the following regulations on the distribution infrastructure?

(Scale 0-10, 0 → no impact, 10 → high adverse impact, N: not sure)

Regulations	**Water Quality Impacts**
Long Term 2 Enhanced Surface Water Treatment Rule (LT2ESWTR)	
Stage 2 Disinfectant/Disinfection byproduct Rule (Stage 2 D/DBPR)	
Groundwater Rule (GWR)	
Total Coliform Rule (TCR)	
Arsenic Rule	
Radon Rule	
Lead and Copper Rule (LCR)	
Other (specify)	

Comments:

9. **What are the major research needs and knowledge gaps in evaluating the impact of water quality on the internal deterioration of the following water distribution material?[a]**

Unlined Iron
Copper
Lead
PVC & PE
Asbestos Cement
Cement Concrete
Linings
Others (specify)

---

[a] Please note down any other research needs that you feel were not addressed in the preceding survey.

REFERENCES

Acero JL, Haderlein SB, Schmidt TC, Suter MJ, and von Gunten U. 2001. MTBE Oxidation by Conventional Ozonation and the Combination Ozone/H_2O_2. *Environmental Science and Technology - Columbus* 35 (21): 4252-4258.

Acero JL, and Gunten Uv. 2001. Characterization of oxidation processes: Ozonation and the AOP O_3/H_2O_2. *American Water Works Association Journal* 93 (10): 90-11.

Acero JL, Rodriguez E, and Meriluoto J. 2005. Kinetics of reactions between chlorine and the cyanobacterial toxins microcystins. *Water research* 39 (8): 1628-11.

Adeloju SB, and Hughes HC. 1986. The corrosion of copper pipes in high chloride - low carbonate mains water. *Corrosion Science* 26 (10): 851-870.

Al-Adeeb AM, and Matti MA. 1984. Leaching corrosion of asbestos cement pipes. *The International Journal of Cement Composites and Lightweight Concrete* 26 (10): 233-240.

Alben, K., Bruchet, A., and Shpirt, E. 1989. *Leachate from organic coating materials used in potable water distribution systems*. Denver, CO. American Water Works Association Research Foundation / American Water Works Association.

Albertsson A-C, Andersson SO, and Karlsson S. 1987. The mechanism of biodegradation of polyethylene. *Polymer Degradation and Stability* 18: 73-87.

Alborzfar M, Jonsson G, and Gron C. 1998. Removal of natural organic matter from two types of humic ground waters by nanofiltration. *Water research* 32 (10): 2983-12.

Anselme C, Mandra V, Baudin I, Jacagelo JG, and Mallevialle J. 1993. Removal of total organic matters and micropollutants by membrane processes in drinking-water treatment. *Water Supply - International Water Supply Association* 11 (3): 249-10.

Appenzeller BMR, Batte M, Mathieu L, Block JC, Lahoussine V, Cavard J, and Gatel D. 2001. Effect of adding phosphate to drinking water on bacterial growth in slightly and highly corroded pipes. *Water research* 35 (4): 1100-1106.

Appenzeller BMR, Duval YB, Thomas F, and Block J-C. 2002. Influence of Phosphate on Bacterial Adhesion onto Iron Oxyhydroxide in Drinking Water. *Environmental Science and Technology - Columbus* 36 (4): 646-647.

AWWA. 2000. Water Quality Division Disinfection Committee Report: Disinfection at large and medium-size systems. *American Water Works Association Journal* 92 (5): 32-43.

AWWA 2001. *Dawn of the replacement era - Reinvesting in drinking water infrastructure*. Denver, CO. American Water Works Association.

AwwaRF, and DVGW TZW. 1996. *Internal Corrosion of Water Distribution Systems: Cooperative Research Report*. Denver, CO: American Water Works Association Research Foundation/ American Water Works Association.

Ayele J, Levavasseur P, and Mazet M. 1996. Triazine adsorption on powdered activated carbon. *Aqua - Oxford* 45 (1): 28-7.

Bade K, Manz W, and Szewzyk U. 2000. Behavior of sulfate reducing bacteria under oligotrophic conditions and oxygen stress in particle-free systems related to drinking water. *FEMS microbiology ecology* 32 (3): 215-10.

Basu DK, Saxena J, Stoss W, Santodonato J, Neal MW, and Kopfler FC. 1987. Comparison of drinking water mutagenicity with leaching polycyclic aromatic hydrocarbons from water distribution systems. *Chemosphere* 16 (10-12): 2595-2612.

Baylis JR. 1926. Factors other than dissolved oxygen influencing corrosion of iron pipes. *Industrial and Engineering Chemistry* 18 (4): 370.

Beardsley M, and Adams CD. 2003. Modeling and Control of Vinyl Chloride in Drinking Water Distribution Systems. *Journal of Environmental Engineering* 129 (9): 844-848.

Benefield LD, and Morgan JM. 1999. Chemical Precipitation. In *Water Quality and Treatment: A Handbook of Community Water Supplies*, ed RD Letterman New York: McGraw Hill.

Berend K, and Trouwborst T. 1999. Cement-mortar pipes as a source of aluminum - Acute aluminum intoxication in dialysis patients was traced to aluminum leached from new cement-mortar pipes. *American Water Works Association Journal* 91 (7): 91-11.

Berend K, Van Der Voet G, and Boer WH. 2001. Acute aluminum encephalopathy in a dialysis center caused by a cement mortar water distribution pipe. *Kidney international* 59 (2): 746-748.

Besner MC, Gauthier V, Barbeau B, Millette R, Chapleau R, and Prevost M. 2001. Understanding distribution system water quality. *American Water Works Association Journal* 93 (7): 101-115.

Bond RG, and DiGiano FA. 2004. Evaluating GAC Performance Using the ICR Database. *Journal of the American Water Works Association* 96 (6): 96-99.

Bonds RW. 2004. Effect of chloramine on ductile-iron pipe gaskets of various elastomer compounds. *Journal of the American Water Works Association* 96 (4): 153-158.

Boulay N, and Edwards M. 2001. Role of temperature, chlorine, and organic matter in copper corrosion by-product release in soft water. *Water research* 35 (3): 683-688.

Boulos PF, Karney BW, Wood DJ, and Lingireddy S. 2005. Hydraulic transient guidelines for protecting water distribution systems. *Journal of the American Water Works Association* 97 (5): 111-114.

Boyd GR, Tarbet NK, Kirmeyer G, Murpby BM, Serpente RE, and Zammit M. 2001. Selecting lead pipe rehabilitation and replacement technologies. *American Water Works Association Journal* 93 (7): 74-14.

Boyd GR, Wang H, Britton MD, Howie DC, Wood DJ, Funk JE, and Friedman MJ. 2004a. Intrusion within a simulated water distribution system due to hydraulic transients. I: description of test rig and chemical tracer method. *Journal of Environmental Engineering* 130 (7): 774-777.

Boyd GR, Wang H, Britton MD, Howie DC, Wood DJ, Funk JE, and Friedman MJ. 2004b. Intrusion within a simulated water distribution system due to hydraulic transients. II: Volumetric method and comparison of results. *Journal of Environmental Engineering* 130 (7): 778-784.

Bremer PJ, Webster BJ, and Wells DB. 2001. Biocorrosion of copper in potable water. *American Water Works Association Journal* 93 (8): 82-10.

Brocca D, Arvin E, and Mosbaek H. 2002. Identification of organic compounds migrating from polyethylene pipelines into drinking water. *Water research* 36 (15): 3675-3676.

Broo AE, Berghult B, and Hedberg T. 1997. Copper corrosion in drinking water distribution systems - The influence of water quality. *Corrosion Science* 39 (6): 1119-1132.

Broo AE, Berghult B, and Hedberg T. 1998. Copper corrosion in water distribution systems - The influence of natural organic matter (NOM) on the solubility of copper corrosion products. *Corrosion Science* 40 (9): 1479-12.

Broo AE, Berghult B, and Hedberg T. 2001a. Drinking water distribution - Improvements of the surface complexation model for iron corrosion. *Water Science and Technology - Water Supply* 1 (3): 11-18.

Broo AE, Berghult B, and Hedberg T. 2001b. Pipe material selection in drinking water systems - A conference summary. *Water Science and Technology - Water Supply* 1 (3): 117.

Butterfield PW, Camper AK, Biederman JA, and Bargmeyer AM. 2002. Minimizing biofilm in the presence of iron oxides and humic substances. *Water research* 36 (15): 3898-13.

Campos C, Schimmoller L, Marinas BJ, Snoeyink VL, Baudin I, and Laine JM. 2000. Adding PAC to remove DOC. *American Water Works Association Journal* 92 (8): 69-15.

Cardew PT. 2000. Simulation of lead compliance data. *Water research* 34 (8): 2241-12.

Carter JT, Rice EW, Buchberger SG, and Lee Y. 2000. Relationships between levels of heterotrophic bacteria and water quality parameters in a drinking water distribution system. *Water research* 34 (5): 1495-1498.

Chen ASC, Fields KA, Sorg TJ, and Wang L. 2002. Field evaluation of As removal by conventional plants. *American Water Works Association Journal* 94 (9): 64-77.

Chen C, Taylor JS, Mulford LA, and Norris CD. 2004. Influences of molecular weight, molecular size, flux, and recovery for aromatic pesticide removal by nanofiltration. *Desalination* 160 (2): 103-112.

Choo K-H, Park I-H, and Choi S-J. 2004. Membranes in drinking water treatment - Removal of natural organic matter using iron oxide-coated membrane systems. *Water Science and Technology - Water Supply* 4 (5-6): 207-208.

Chwirka J, Colvin C, Gomez JD, and Mueller PA. 2004. Arsenic removal from drinking water using coagulation microfiltration process. *American Water Works Association Journal* 96 (3): 106-113.

Clark RM, Adams JQ, Sethi V, and Sivagaesan M. 1998. Control of microbial contaminants and disfection by-products for drinking water in the US: Cost and performance. *Aqua - Oxford* 47 (6): 255-11.

Cleasby JL, and Logsdon GS. 1999. Granular and precoat filtration. In *Water Quality and Treatment: A Handbook of Community Water Supplies*, ed RD Letterman New York: McGraw Hills.

Clifford DA. 1999. Ion Exchange and inorganic adsorption. In *Water Quality and Treatment: A Handbook of Community Water Supplies*, ed RD Letterman New York: McGraw Hills.

Clifford DA, Ghurye GL, and Tripp AR. 2003. As removal using ion exchange with spent brine recycling. *American Water Works Association Journal* 95 (6): 119-130.

Colombo AF, and Karney BW. 2002. Energy and costs of leaky pipes: Towards comprehensive picture. *Journal of Water Resources Planning and Management* 128 (6): 441-450.

Craun GF, Calderon RL, and Craun MF. 2003. Waterborne outbreaks in the United States. In *Drinking Water Regulation and Health*, ed FW Pontius New York: Wiley-Interscience.

Critchley MM, Cromar NJ, Mcclure NC, and Fallowfield HJ. 2003. The influence of the chemical composition of drinking water on cuprosolvency by biofilm bacteria. *Journal of applied microbiology* 94 (3): 501-507.

Critchley MM, Pasetto R, and O'Halloran RJ. 2004. Microbiological influences in 'blue water' copper corrosion. *Journal of applied microbiology* 97 (3): 590-598.

Cromwell III JE, Reynolds H, Pearson J, and Grant M. 2002. *Costs of Infrastructure Failure.* Denver, CO: AwwaRF.

Daniel PA. 1998. *Balancing Multiple Water Quality Objectives.* Denver, CO: AwwaRF.

Dartmann J, Alex T, Dorsch T, Schevalje E, and Johannsen K. 2004. Influence of Decarbonisation and Phosphate Dosage on Copper Corrosion in Drinking Water Systems. *Acta Hydrochimica et Hydrobiologica* 32 (1): 25-28.

Davies C, Kaucner C, Altavilla N, Ashbolt N, Ferguson C, Krogh M, Hijnen W, Medema G, and Deere D. 2005. *Fate and Transport of Surface Water Pathogens in Watersheds*. Denver, CO: AwwaRF.

Deb AK, Grabultz FM, Hasit YJ, and Momberger KA. 2000. *Guidance for Management of Distribution System Operation and Maintenance*. Denver, CO: AwwaRF.

Diehl AC, Speitel GE, Symons JM, Krasner SW, Hwang CJ, and Barrett K. 2000. DBP formation during chloramination. *American Water Works Association Journal* 92 (6): 76-93.

Dorn, R., Houseman, P., Hyde, R. A., and Jarvis, M. G. 1996. *Water Mains: Guidance on Assessment and Inspection Techniques*.London, UK. Construction Industry Research and Information Association.

Douglas BD, Merrill DT, and Catlin JO. 1996. Water quality deterioration from corrosion of cement-mortar linings. *American Water Works Association Journal* 88 (7): 99.

Edwards M, Hidmi L, and Gladwell D. 2002. Phosphate inhibition of soluble copper corrosion by-product release. *Corrosion Science* 44: 1057-1071.

Edwards M, and Sprague N. 2001. Organic matter and copper corrosion by-product release: A mechanistic study. *Corrosion Science* 43 (1): 1-18.

Edwards M, and Dudi A. 2004. Role of chlorine and chloramine in corrosion of lead-bearing plumbing materials. *Journal of the American Water Works Association* 96 (10): 69-14.

Edwards M, Jacobs S, and Taylor RJ. 2000. The blue water phenomenon. *American Water Works Association Journal* 92 (7): 72-11.

Edwards M, and McNeill LS. 2002. Effect of phosphate inhibitors on lead release from pipes. *American Water Works Association Journal* 94 (1): 79-12.

Edwards M, Patel S, McNeill L, Chen Hw, Frey M, Eaton AD, Antweiler RC, and Taylor HE. 1998. Considerations in As analysis and speciation. *American Water Works Association Journal* 90 (3): 103-113.

Edwards M, Schock MR, and Meyer TE. 1996. Alkalinity, pH, and copper corrosion by-product release. *American Water Works Association Journal* 88 (3): 81-14.

Fischer WR, Wagner D, Siedlarek H, and Sequeira JAGC. 1993. Perforations of polypropylene pipes in potable water caused by cracking - A case study. 8 1992, at Great Britain.

Fisher EL, Fuortes LJ, Valentine RL, Mehrhoff M, and Field RW. 2000. Dissolution of 226 Radium from pipe-scale deposits in a public water supply. *Environment International - a Journal of Science Technology Health Monitoring and Policy* 26 (1-2): 69-6.

Fransolet G, Depelchin A, Villers G, Goossens R, and Masschelein WJ. 1999. The role of bicarbonate in bacterial growth in oligotrophic waters. *Water research* 39 (9): 1809-1817.

Frateur L, Deslouis C, Kiene L, Levi Y, and Tribollet B. 1999. Free chlorine consumption induced by cast iron corrosion in drinking water distribution systems. *Water research* 33 (8): 1781-10.

Gagnon GA, Rand JL, leary KC, Rygel AC, Chauret C, and Andrews RC. 2005. Disinfectant efficacy of chlorite and chlorine dioxide in drinking water biofilms. *Water research* 39 (9): 1809.

Ghorai S, and Pant KK. 2004. Investigations on the column performance of Fluoride adsorption by activated alumina in a fixed-bed. *Chemical Engineering Journal* 98 (1): 165-174.

Ghorai S, and Pant KK. 2005. Equilibrium, kinetics and breakthrough studies for adsorption of Fluoride on activated alumina. *Separation and Purification Technology* 42 (3): 265-272.

Ghurye GL, and Clifford DA. 2004. As(III) oxidation using chemical and solid-phase oxidants. *American Water Works Association Journal* 96 (1): 84-96.

Ghurye GL, Clifford DA, and Tripp AR. 1999. Combined Arsenic and Nitrate removal by ion exchange. *American Water Works Association Journal* 91 (10): 85-96.

Ghurye GL, Clifford DA, and Tripp AR. 2004. Iron coagulation and direct microfiltration to remove Arsenic from groundwater. *American Water Works Association Journal* 96 (4): 143-152.

Gill TS, Knapp RJ, Bradley SW, and Bradley WL. 1999. Long term durability of crosslinked polyethylene tubing used in chlorinated hot water systems. *Plastics, Rubber and Composites* 28 (6): 309-313.

Gregor J. 2001. Arsenic removal during conventional aluminium-based drinking-water treatment. *Water research* 35 (7): 1659-6.

Griffini O, Bao ML, Burrini D, Santianni D, Barbieri C, and Pantani F. 1999. Removal of pesticides during the drinking water treatment process at Florence water supply, Italy. *Aqua - Colchester* 48 (5): 177-179.

Grigg NS. 2005. Assessment and renewal of water distribution systems. *American Water Works Association Journal* 97 (2): 58-68.

Guo Q, Toomuluri PJ, and Eckert JO. 1998. Leachability of regulated metals from cement-mortar linings - In a laboratory study, metals leached from cement-mortar linings at levels that suggest water suppliers should be selective when choosing cement for lining pipes. *American Water Works Association Journal* 90 (3): 62-12.

Gurian PL, and Small MJ. 2002. Point-of-use Treatment and the Revised Arsenic MCL. *Journal AWWA* 94 (3): 101-108.

Haas CN. 1999a. Benefits of using a disinfectant residual. *American Water Works Association Journal* 91 (1): 65-69.

Haas CN. 1999b. Disinfection. In *Water Quality and Treatment: A Handbook of Community Water Supplies*, ed RD Letterman New York: McGraw Hill.

Haas CN, Joffe J, Heath M, Jacangelo J, and Anmangandla U. 1998. Predicting disinfection performance in continuous flow systems from batch disinfection kinetics. *Water Science and Technology* 38 (6): 171-10.

Hallam NB, West JR, Forster CF, Powell JC, and Spencer I. 2002. The decay of chlorine associated with the pipe wall in water distribution systems. *Water research* 36 (14): 3479-10.

Han B, Runnells TR, Zimbron TR, and Wickramasinghe SR. 2002. Arsenic removal drom drinking water by flocculation and microfiltration. *Desalination* 145 (1-3): 293-298.

Han B, Zimbron TR, Runnells TR, Shen Z, and Wickramasinghe SR. 2003. New Arsenic Standard spurs search for cost-effective removal techniques. *American Water Works Association Journal* 95 (10): 109-118.

Hand DW, Hokanson DR, and Crittenden JC. 1999. Air Stripping and Aeration. In *Water Quality and Treatment: A Handbook of Community Water Supplies*, ed RD Letterman New York: McGraw Hill.

Harrison DB, Nicholas DM, and Evans GM. 2004. Pitting corrosion of copper tubes in soft drinking waters: Corrosion mechanism. *Journal of the American Water Works Association* 96 (11): 67-10.

Hassinen J, Lundback M, Ifwarson M, and Gedde UW. 2004. Deterioration of polyethylene pipes exposed to chlorinated water. *Polymer Degradation and Stability* 84: 261-267.

Holden B, Greetham M, Croll BT, and Scutt J. 1995. The effect of changing inter process and final disinfection reagents on corrosion and biofilm growth in distribution pipes. *Water Science and Technology* 32 (8): 213-220.

Hong PKA, and Macauley Y-Y. 1998. Corrosion and Leaching of Copper Tubing Exposed to Chlorinated Drinking Water. *Water Air and Soil Pollution* 108 (3): 457-16.

Imran SA. 1997. Removal of microorganisms through slow sand filtration of chlorinated vs. non-chlorinated secondary effluents. M.S. Removal of microorganisms through slow sand filtration of chlorinated vs. non-chlorinated secondary effluents, King Fahd University of Petroleum & Minerals.

Imran SA, Dietz JD, Mutoti G, Taylor JS, and Randall AA. 2005a. Modified Larsons Ratio incorporating temperature, water age, and electroneutrality effects on red water release. *Journal of Environmental Engineering* 131 (11): 1514-1517.

Imran SA, Dietz JD, Mutoti G, Xiao W, Taylor JS, and Desai V. 2006. Optimizing source water blends for corrosion and residual control in distribution systems. *American Water Works Association Journal* 98 (5): 107-115.

Imran SA, Dietz JD, Mutoti G, Taylor JS, Randall AA, and Cooper CD. 2005b. Red water release in drinking water distribution systems. *Journal of the American Water Works Association* 97 (9): 93-10.

Isaac RA, Gil L, Cooperman AN, Hulme K, Eddy B, Ruiz M, Jacobson K, Larson C, and Pancorbo OC. 1997. Corrosion in drinking water distribution systems: A major contributor of Copper and Lead to wastewaters and effluents. *Environmental Science and Technology - Columbus* 31 (11): 3198-6.

Jacobs S, Reiber S, and Edwards M. 1998. Sulfide-induced copper corrosion. *American Water Works Association Journal* 90 (7): 62-12.

Karanfil T, and Kilduff JE. 1999. Role of granular activated carbon surface chemistry on the adsorption of organic compounds. 1. Priority Pollutants. *Environmental Science and Technology - Columbus* 33 (18): 3217-3218.

Karim MR, Abbaszadegan M, and LeChevallier M. 2003. Potential for pathogen intrusion during pressure transients. *American Water Works Association Journal* 95 (5): 134-13.

Kilb B, Lange B, Schaule G, Flemming H, and Wingender J. 2003. Contamination of drinking water by Coliforms from biofilms growth on rubber-coated valves. *International journal of hygiene and environmental health* 206 (6): 563-573.

Kim J, Benjamin MM, Kwan P, and Chang Y. 2003. A novel ion exchange process for As removal. *American Water Works Association Journal* 95 (3): 77-85.

Kirisits MJ, Snoeyink VL, Chee-sanford JC, Daugherty BJ, Brown JC, and Raskin L. 2002. Effect of operating conditions on bromate removal efficiency in BAC filters. *American Water Works Association Journal* 94 (4): 182-193.

Kirisits MJ, Snoeyink VL, Inan H, Chee-sanford JC, Raskin L, and Brown JC. 2001. Water quality factors affecting bromate reduction in biologically active carbon filters. *Water research* 35 (4): 891-900.

Kirmeyer, Gregory J., Richards, W., and Smith, C. D. 1994. *An assessment of water distribution systems and associated research needs*. Denver, CO. American Water Works Association Research Foundation / American Water Works Association.

Kunzru S, and Chaudhuri M. 2005. Manganese amended Activated Alumina for adsorption/oxidation of Arsenic. *Journal of Environmental Engineering* 131 (9): 1350-1353.

Lagos GE, Cuadrado CA, and Letelier MV. 2001. Aging of copper pipes by drinking water. *American Water Works Association Journal* 93 (11): 94-10.

Larson TE, and Skold RV. 1957. Corrosion and tuberculation of cast iron. *American Water Works Association Journal*: 1294-1302.

Lau BLT, Harrington GW, Anderson MA, and Tejedor I. 2005. Physicochemical aspects of Cryptosporidium surrogate removal in carbon block filtration. *Journal of American Water Works Association* 97 (2): 92-102.

LeChevallier MW. 1999. The case for maintaining a disinfectant residual. *American Water Works Association Journal* 91 (1): 86-89.

Lee S-M, Choi B-J, Kim D-H, Lee I-Y, and Kim W. 2003. Selected - Competitive adsorption for binary mixture of 4-nitrophenol and phenol on RSTA using GAC. *Water Science and Technology* 47 (1): 229-8.

Lehtola MJ, Miettinen I, Keinanen MM, Kekki TK, Laine O, Hirvonen A, Vartiainen T, and Martikainen PJ. 2004a. Microbiology, chemistry and biofilm development in a pilot drinking water distribution system with copper and plastic pipes. *Water research* 38: 3769-3779.

Lehtola MJ, Miettinen IT, Lampola T, Hirvonen A, Vartiainen T, and Martikainen PJ. 2005b. Pipeline materials modify the effectiveness of disinfectants in drinking water distribution systems. *Water research* 39 (10): 1962-10.

Lehtola MJ, Nissinen TK, Miettinen IT, Martikainen PJ, and Vartiainen T. 2004b. Removal of soft deposits from the distribution system improves the drinking water quality. *Water research* 38 (3): 601-610.

Leiknes T, Odegaard H, and Myklebust H. 2004. Removal of natural organic matter (NOM) in drinking water treatment by coagulation-microfiltration using metal membranes. *Journal of Membrane Science* 242 (1): 47-10.

Leroy, P., Schock, M. R., Wagner, I., and Holtschulte, H. 1996. *Cement-based materials*. Denver, CO. American Water Works Association Research Foundation/ American Water Works Association.

Letterman RD, Amritrajah A, and O'Melia CR. 1999. Coagulation and Flocculation. In *Water Quality and Treatment: A Handbook of Community Water Supplies*, ed RD Letterman New York: McGraw Hill.

Lin J, Ellaway M, and Adrien R. 2001. Study of corrosion material accumulated on the inner wall of steel water pipe. *Corrosion Science* 43 (11): 2065-18.

Lin N-H, Torrents A, Davis AP, Zeinali M, and Taylor FA. 1997. Lead Corrosion Control from Lead, Copper-Lead Solder, and Brass Coupons in Drinking Water Employing Free and Combined Chlorine. *Journal of Environmental Science and Health - Part A - Environmental Science Engineering* 32 (4): 865-20.

Little B, and Ray R. 2002. A Perspective on Corrosion Inhibition by Biofilms. *Corrosion - The Journal of Science and Engineering* 58 (5): 424-429.

Little B, Wagner P, and Mansfield F. 1992. An Overview of Microbiologically Influenced Corrosion. *Electrochimica Acta* 37 (12): 2185-2194.

Liu S, Taylor JS, Randall AA, and Dietz JD. 2005. Nitrification modeling in chloraminated distribution systems. *Journal of the American Water Works Association* 97 (10): 98-11.

Lovins WA, III, Duranceau SJ, Powell RM, and Voorhees JR. 2005. Experiences with blending multiple source waters in a common water distribution systems. *Florida Water Resources Journal*, 2005, 31-35.

Lundback M, Hassinen J, Andersson U, Fujiwara T, and Gedde UW. 2005. Polybutene-1 pipes exposed to pressurized chlorinated water: Lifetime and antioxidant consumption. *Polymer Degradation and Stability* XXX (XXX): 1-6.

Lytle DA, Sorg TJ, and Frietch C. 2004. Accumulation of Arsenic in Drinking Water Distribution Systems. *Environmental Science and Technology - Columbus* 38 (20): 5365-5368.

Lytle DA, Gerke TL, and Maynard JB. 2005. Effect of bacterial sulfate reduction on iron-corrosion scales. *Journal of the American Water Works Association* 97 (10): 109-116.

Lytle DA, Schock MR, Clement JA, and Spencer CM. 1998. Using aeration for corrosion control. *American Water Works Association Journal* 90 (3): 74-15.

Magara Y, Aizawa T, Kunikane S, Itoh M, Kohki M, Kawasaki M, and Takeuti H. 1996. The behavior of inorganic constituents and disinfection by products in reverse osmosis water desalination process. *Water Science and Technology* 34 (9): 141-148.

Maier M, Maier D, and Lloyd BJ. 1999. The mobilisation of polycyclic aromatic hydrocarbons (PAHs) from the coal-tar lining of water pipes. *Aqua - Colchester* 48 (6): 238-12.

Maier M, Maier D, and Lloyd BJ. 2000. Factors influencing the mobilisation of polycyclic aromatic hydrocarbons (PAHs) from the coal-tar lining of water mains. *Water research* 34 (3): 773-14.

Marangou VS, and Savvides K. 2001. First Desalination Plant in Cyrus - Product Water Aggresivity and Corrosion Control. *Desalination* 138 (1-3): 251-258.

Marchesan M, and Morran J. 2004. Tastes associated with products in contact with drinking water. *Water Science and Technology* 49 (9): 227-6.

Matilainen A, Liikanen R, Nystrom M, Lindqvist N, and Tuhkanen T. 2004. Enhancement of the Natural Organic Matter Removal from Drinking Water by Nanofiltration. *Environmental Technology - London* 25 (3): 283-10.

McNeill LS, and Edwards M. 2001. Iron pipe corrosion in distribution systems. *American Water Works Association Journal* 93 (7): 88-13.

McNeill LS, and Edwards M. 2002. The Importance of Temperature in Assessing Iron Pipe Corrosion in Water Distribution Systems. *Environmental monitoring and assessment* 77 (3): 229-14.

Mehta PK. 2003. *Concrete in the marine environment*. Essex, England: Elsevier Science Publishers Ltd.

Melo LF, and Vieira MJ. 1999. Physical Stability and Biological Activity of Biofilms Under Turbulent Flow and Low Substrate Concentration. *Bioprocess Engineering* 20 (4): 363-368.

Merkel T, Maier M, Sacher F, and Maier D. 1997. Reactions of polynuclear aromatic hydrocarbons with chorine and chlorine dioxide in coal tar lined pipes. *Aqua - Oxford* 46 (6): 289-15.

Merkel TH, Gross H, Dahlke T, Reicherter S, Beuchle G, and Eberle SH. 2002. Copper Corrosion by-Product Release in Long-Termstagnation Experiments. *Water research* 36: 1547-1555.

Morin P, Gauthier V, Saby S, and Block J-C. 1999. Bacterial Resistance to Chlorine through Attachment to Particles and Pipe Surfaces in Drinking Water Distribution Systems. *Special Publications of the Royal Society of Chemistry*: 171-20.

Narasimhan, R. 2005. *Management of disposal of radioactive residuals in drinking water treatment.*Denver, CO. AwwaRF.

Narbaitz RM, Mayorga WJ, Torres P, Greenfield JH, Amy GL, and Minear RA. 2002. Evaluating Aeration-Stripping Media on the Pilot Scale. *Journal of American Water Works Association* 94 (9): 97-111.

Neuberger M, Frank W, Golob P, and Warbichler P. 1995. Asbestos Concentrations in Drinking Water Asbestos Cement Pipes and Geogen Sources in Austria. *Zentralblatt fur Hygiene und Umweltmedizin* 198 (4): 293-14.

Ning RY. 2002. Arsenic Removal by Reverse Osmosis. *Desalination* 143 (3): 237-241.

Oakes K. 2005. Membrane Filtration Processes for Removing Arsenic from Drinking Water. 15 2005.

Palit A, and Pehkonen SO. 2000. Copper Corrosion in Distribution Systems: Evaluation of a Homogeneous Cu_2O Film and a Natural Corrosion Scale Inhibitors. *Corrosion Science* 42: 1801-1822.

Paune F, Caixach J, Espadaler I, Om J, and Rivera J. 1998. Assessment on the removal of organic chemicals from raw and drinking water at a Llobregat river water works plant using GAC. *Water research* 32 (11): 3313-12.

Peters NJ, Davidson CM, Britton A, and Robertson SJ. 1999. The nature of corrosion products in lead pipes used to supply drinking water to the City of Glasgow, Scotland, UK. *Fresenius Journal of Analytical Chemistry* 363 (5): 562-564.

Pontius FW. 1991. Phase II organic and inorganic contaminant regulations. *American Water Works Association Journal* 83 (8): 20-22-77 - 79.

Pontius FW. 2003a. *Drinking Water Regulations and Health*. New York: Wiley-Interscience.

Pontius FW. 2003b. Update on USEPA's drinking water regulations. *American Water Works Association Journal* 95 (3): 57-12.

Pontius FW. 2004. Drinking water contaminant regulation -- Where are we heading? *Journal of the American Water Works Association* 96 (3): 56-14.

Pyrzynska K, Drzewicz P, and Trojanowicz M. 1998. Preconcentration and Separation of Inorganic Selenium Species on Activated Alumina. *Analytica Chimica Acta - Including Cumulative Indexes* 363 (2): 141-146.

Rajani, B2006b *Personal Communication*

Rajani, B2006a *Personal Communication*

Roberson JA. 2003. Complexities of the new drinking water regulations -- Everything you wanted to know but were afraid to ask. *American Water Works Association Journal* 95 (3): 48-49.

Rosenfeldt EJ, and Linden KG. 2004. Degradation of Endocrine Disrupting Chemicals Bisphenol A, Ethinyl Estradiol, and Estradiol during UV Photolysis and Advanced Oxidation Processes. *Environmental Science and Technology - Columbus* 38 (20): 5476-5478.

Rushing JC, and Edwards M. 2004. The Role of Temperature Gradients in Residential Copper Pipe Corrosion. *Corrosion Science* 46: 1883-1894.

Rushing JC, McNeil LS, and Edwards M. 2003. Some Effects of Aqueous Silica on the Corrosion of Iron. *Water research* 37 (5): 1080-1090.

Sadiq R, Kleiner Y, and Rajani B. 2004. Aggregative risk analysis for water quality failure in distribution networks. *Journal of Water Supply: Research & Technology - AQUA* 53 (4): 241-261.

Sadiq R, Kleiner Y, and Rajani B. 2006. Estimating risk of contaminant intrusion in distribution networks using Dempster-Shafer theory of evidence. *Civil Engineering and Environmental Systems* 23 (3): 129-141.

Sarin P, Snoeyink VL, Bebee J, Jim KK, Beckett MA, Kriven WM, and Clement JA. 2004a. Iron release from corroded iron pipes in drinking water distribution systems: effect of dissolved oxygen. *Water research* 38 (5): 1259-11.

Sarin P, Snoeyink VL, Bebee J, Kriven WM, and Clement JA. 2001. Physico-chemical characteristics of corrosion scales in old iron pipes. *Water research* 35 (12): 2961-2969.

Sarin P, Snoeyink VL, Lytle DA, and Kriven WM. 2004b. Iron corrosion scales: Model fro scale growth, iron release and colored water formation. *Journal of Environmental Engineering* 130 (4): 364-373.

Sarin P, Clement JA, Snoeyink VL, and Kriven WW. 2003. Iron release from corroded unlined cast-iron pipe. *American Water Works Association Journal* 95 (11): 85-12.

Satchwill T. 1998. "The impact of pipe coatings on drinking water quality." 2002.

Scharfenaker MA. 2002. White Papers Set Stage for Regulating Distribution System Water Quality. *Journal of American Water Works Association*: 14-24.

Schock MR. 1999. Internal corrosion and deposition control. ed RD Letterman New York: McGraw Hill.

Schock MR, Holldber J, Lovejoy TR, and Lowry J. 2002. California's First Aeration Plants for Corrosion Contro. *Journal of American Water Works Association* 94 (3): 88-100.

Schock, M. R., Lytle, D. A., and Clement, J. A. 1995. *Effect of pH, DIC, Orthophosphate and Sulfate on Drinking Water Cuprosolvency*.Cincinnatti, Ohio. USEPA.

Schock, M. R., Wagner, I., and Oliphant, Roger J. 1996. *Corrosion and solubility of lead in drinking water*.Denver,CO. American Water Works Association Research Foundation/ American Water Works Association.

Shim JJ, and Kim JG. 2004. Copper Corrosion in Potable Water Distribtion Systems: Influence of Copper Products on the Corrosion Behaviour. *Materials Letters* 58 (14): 2002-2006.

Slaats PGG, Mesman GAM, Rosenthal LPM, and Brink H. 2004. Tools to monitor corrosion of cement-containing water mains. *Water Science and Technology* 49 (2): 33-38.

Smith DB, Clark RM, Pierce BK, and Regli S. 1995. An empirical model for interpolating C . t values for chlorine inactivation of Giardia lamblia. *Aqua - Oxford* 44 (5): 203-209.

Snoeyink VL, and Summers RS. 1999. Adsorption of organic compounds. In *Water Quality and Treatment: A Handbook of Community Water Supplies*, ed RD Letterman New York: McGraw Hill.

Sorg TJ, and Logsdon GS. 1980. Treatment Technology to Meet the Interim Primary Drinking Water Regulations for Inorganics: Part 5. *Journal of American Water Works Association* 72 (7): 411-422.

Sorrell RK, Brass HJ, and Reding R. 1980. A Review of Occurrences and Treatment of Polynuclear Aromatic Hydrocarbons in Water. *Environment International* 4 (3): 245-254.

Springthorpe S, Sander M, Nolan K, and Sattar SA. 2001. Comparison of static and dynamic disinfection models for bacteria and viruses in water of varying quality. *Water Science and Technology* 43 (12): 147-148.

Sutherland D, Swash PM, Macqueen AC, McWilliam LE, Ross DJ, and Wood SC. 2002. A Field Based Evaluation of Household Arsenic Removal Technologies for the Treatment of Drinking Water. *Environmental Technology - London* 23 (12): 1385-20.

Tang Z, Hong SK, Xiao W, and Taylor J. 2005. Characteristics of iron corrosion scales established under blending of ground, surface, and saline waters and their impacts on iron release in the pipe distribution system. *Corrosion Science* xx (xx): xx.

Taylor JS, Dietz JD, Randall AA, Hong SK, Norris CD, Mulford LA, Arevalo JM, Imran S, Lepuil M, Mutoti I, Tang J, Xiao W, Cullen C, Heaviside R, Mehta A, Patel M, Vasquez F, and Webb D. 2005. *Effects of Blending on Distribution System Water Quality*. Denver, CO.: American Water Works Research Foundation.

Taylor JS, and Wiesner M. 1999. Membranes. In *Water Quality and Treatment: A Handbook of Community Water Supplies*, ed RD Letterman New York: McGraw Hill.

Thirunavukkarasu OS, Viraraghavan T, Subramanian KS, Chaalal O, and Islam MR. 2005. Arsenic Removal in Drinking Water--Impacts and Novel Removal Technologies. *Energy Sources - Journal of Extraction Conversion and the Environment* 27 (1): 209-212.

Thurston-Enriquez JA, Haas CN, Jacangelo J, and Gerba CP. 2003. Chlorine Inactivation of Adenovirus Type 40 and Feline Calicivirus. *Applied and Environmental Microbiology* 69 (7): 3979-7.

Tomboulian P, Schweitzer L, Mullin K, Wilson J, and Khiari D. 2004. Materials used in drinking water distribution systems: Contribution to taste-and-odor. *Water Science and Technology* 49 (9): 219-8.

Torvinen E, Suomalainen S, Lehtola MJ, Miettinen IT, Zacheus O, Paulin L, Katila ML, and Martikainen PJ. 2004. Mycobacteria in Water and Loose Deposits of Drinking Water Distribution Systems in Finland. *Applied and Environmental Microbiology* 70 (4): 1973-1979.

Trussell, R. R. and Wagner, I. 1996. *Corrosion of galvanized pipe*.Denver, CO. American Water Works Association Research Foundation/ American Water Works Association.

USEPA 1998a. *Small System Compliance Technology List for the Non-Microbial Contaminants Regulated before 1996*.United States Environmental Protection Agency.

USEPA 1998b. *Small System Compliance Technology List for the Surface Water Treatment Rule and Total Coliform Rule*.United States Environmental Protection Agency.

USEPA 1999a. *Alternative Disinfectants and Oxidants Guidance Manual*.United States Environmental Protection Agency.

USEPA 1999b. *Microbial and Disinfection Byproduct Rules Simultaneous Compliance Guidance Manual*.United States Environmental Protection Agency.

USEPA 2001a. *Controlling Disinfection Byproducts and Microbial Contaminants in Drinking Water*.United States Environmental Protection Agency.

USEPA 2001b. *Drinking Water Infrastructure Needs Survey - Second Report to Congress*.United States Environmental Protection Agency.

USEPA 2003. *Arsenic Treatment Technology Evaluation Handbook for Small Systems*.United States Environmental Protection Agency.

USEPA 2004. *Providing Safe Drinking Water in America - 2002 National Public Water Systems Compliance Report*.United States Environmental Protection Agency.

Valentine RL, and Stearns SW. 1994. Radon Release from Water Distribution System Deposits. *Environmental Science & Technology* 28 (3): 534-537.

Van der Bruggen B, and Vandecasteele C. 2003. Removal of pollutants from surface water and groundwater by nanofiltration: overview of possible applications in the drinking water industry. *Environmental Pollution - Kidlington* 122 (3): 435-12.

Vasquez FA, Heaviside R, Tang Z, and Taylor JS. 2006. Effect of Free Chlorine and Chloramines on Lead Release in a Distribution System. *Journal of American Water Works Association* 98 (2): 144-154.

Vatankhah G, Drogowska M, Menard H, and Brossard L. 1998. Electrodissolution of Iron in Sodium Sulfate and Sodium Bicarbonate Solutions at pH 8. *Journal of Applied Electrochemistry* 28: 173-183.

Vikesland PJ, and Valentine RL. 2002a. Iron Oxide Surface-Catalyzed Oxidation of Ferrous Iron by Monochloramine: Implications of Oxide Type and Carbonate on Reactivity. *Environmental Science and Technology - Columbus* 36 (3): 512.

Vikesland PJ, and Valentine RL. 2002b. Modeling the Kinetics of Ferrous Iron Oxidation by Monochloramine. *Environmental Science and Technology - Columbus* 36 (4): 662-667.

Vikesland, P. J., Valentine, R. L., Angerman, B. D., Hackett, S. A., Shoup, M., and Slattenlow, S. 2000. *The role of pipe-water interface in DBP formation and disinfectant loss*.Denver, Colorado. AwwaRF and AWWA.

Vikesland PJ, and Valentine RL. 2000. Reaction Pathways Involved in the Reduction of Monochloramine by Ferrous Iron. *Environmental Science and Technology - Columbus* 34 (1): 83-88.

Wang L, Chen ASC, Sorg TJ, and Fields KA. 2002. Field Evaluation of Arsenic Removal by IX and AA. *Journal of American Water Works Association* 94 (4): 161-173.

ABBREVATIONS

AC	Asbestos Cement
AMCL	Alternate Maximum Contaminant Level
AWWA	American Water Works Association
AwwaRF	Awwa Research Foundation
DSM	Distribution System Material
GWR	Groundwater Rule
GWUDI	Groundwater under direct influence of surface waters
IESWTR	Interim Enhanced Surface Water Treatment Rule
LCR	Lead and Copper Rule
LT 1 ESWTR	Long Term 1 Enhanced Surface Water Treatment Rule
LT 2 ESWTR	Long Term 2 Enhanced Surface Water Treatment Rule
MCL	Maximum Contaminant Level
MCLG	Maximum Contaminant Level Goals
MRDL	Maximum Disinfectant Residual Level
MRDLG	Maximum Disinfectant Residual Level Goal
NIPDWR	National Interim Primary Drinking Water Regulation
NPDWR	National Primary Drinking Water Regulation
NRCC	National Research Council of Canada
Regs.	Water Quality Regulations
SDWA	Safe Drinking Water Act
Stage 1 D/DBPR	Stage 1 Disinfectant / Disinfection Byproduct Rule
Stage 2 D/DBPR	Stage 2 Disinfectant / Disinfection Byproduct Rule
SWTR	Surface Water Treatment Rule
T&O	Taste and Odor
TCR	Total Coliform Rule
TT	Treatment Technique
USEPA	United States Environmental Protection Agency
UT	Unit Technology (Treatment Practice)
WQP	Water Quality Parameter

IWA Publishing's authorised EU representative for General Product Safety Regulations is Diane D'Arras, 15 rue Duret, 75116 Paris, France, e-mail: safety@iwap.co.uk.

Printed and bound by CPI Group (UK) Ltd, Croydon, CR0 4YY
17/09/2026
02218811-0008